# 晚清郵驛之演變

POSTAL COMMUNICATION IN CHINA AND ITS MODERNIZATION,
1860-1896

by

Ying-wan Cheng

虹橋書店
Rainbow-Bridge Book Co.

The East Asian Research Center at Harvard
University administers research projects designed
to further scholarly understanding of China, Japan,
Korea, Vietnam, and adjacent areas. These studies
have been assisted by grants from the Ford Foundation.

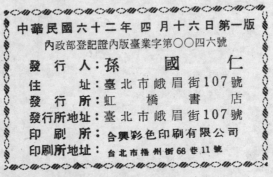

中華民國六十二年 四 月 十 六 日 第一版
內政部登記證內版臺業字第○○四六號
發 行 人：孫 國 仁
住　　　址：臺 北 市 峨 眉 街107號
發 行 所：虹 橋 書 店
發行所地址：臺 北 市 峨 眉 街107號
印 刷 所：合興彩色印刷有限公司
印刷所地址：台北市梧州街68巷11號

Price N.T. $35.00 Net

Library of Congress No. 70-120316
SBN 674-69320-5

*To my parents*

# FOREWORD

Dr. Ying-wan Cheng's case study of the origin of the modern Chinese post office illustrates a number of themes that characterized China's struggle for modernization in the late nineteenth century. First of all, there was a well established old order which met the needs of the Chinese state and people on a pre-modern basis. The state had its long established official post, a far flung network of post stations on major routes over which horse relays carried couriers with official communications, as well as personages, and goods. Among the urban populace there had grown up a system of private letter companies or hongs which sent mail to distant places and charged fees according to the distance and difficulty of the service. Because so much of the society's needs could be handled under this well developed old system, there was less incentive for the modern kind of centralized and pervasive postal service.

A second characteristic feature of the modernization problem was the presence of the foreigner. Under extraterritoriality foreign post offices were established in the treaty ports to serve the foreign community. These were set up on a national basis in each case—British, American, Japanese, et al.— and expanded as the foreign element in the treaty ports expanded, beyond Chinese control. In this way the modernizing sector of China in the treaty ports saw a foreign growth distinct from the old order.

In the third place there was a Chinese response to the stimulus of foreign example in the treaty ports. As trade increased, the needs of the Chinese merchant community caused a proliferation of the traditional private letter hongs that sent mail in the old unofficial way. In other words, the impetus for modernization was by no means confined to the foreigner, but also appeared among the Chinese commercial class.

Finally, the creation of the modern Chinese post office was achieved through the channels of the Chinese government but with distinct foreign assistance. In this case the foreign element was represented chiefly by Robert Hart, Inspector General of the Imperial Maritime Customs Service. He early used his commissioners of Customs at the treaty ports to inaugurate postal services. By patient effort decade after decade he overcame the vested interests that stood in the way of the modern post office. Who can say whet' er Hart and the Customs were "foreign" or "Chinese"? The Customs collected China's revenues on foreign trade under the direction of the Tsungli Yamen at Peking. The foreign commissioners of Customs were part of the Chinese bureaucracy. Only after the disasters of 1895 and 1900 did the Customs have to become an agent for paying foreign bond holders and indemnities, and even then it remained an agent of the Chinese government.

Here we see the anomaly of China's late nineteenth century efforts at modernization: foreignizing elements within the Chinese government tried to foster modernization, principally by imitation of the West, but since this was a period of foreign aggression pursued within the framework of the unequal treaties, there was an inevitable Chinese distaste for all things foreign. Thus the vested interests of China's old order were reenforced by a proto-nationalistic spirit of noncooperation and anti-Westernization.

Summary characterizations of this sort can be put forth at this time only as hypotheses, not as proved generalizations. We are still in the process of building up the body of case studies upon which such generalizations may rest. Dr. Cheng is herself well qualified to contribute to this growing corpus of studies. Her father was a judge on the Permanent Court of International Justice and served as Chinese ambassador to Britain 1946-1950, and her early life abroad gave her an intimate view of the problems of adjustment between China and the West. The present volume has been developed from her Ph.D. dissertation at Radcliffe College.

John K. Fairbank
May 1970

## ACKNOWLEDGEMENTS

This study was originally presented to Radcliffe College in 1959 as a Ph.D. dissertation. Much of the writing was done under the auspices of the East Asian Research Center at Harvard University as one of its projects in Chinese Political and Economic Studies. In revising the work for publication, I have been indebted to the American Association of University Women for a post-doctoral fellowship and to Dowling College (formerly Adelphi Suffolk College of Adelphi University) for a leave of absence in 1965-66. I very much regret that circumstances have unduly delayed its appearance in print before this time.

Among the individuals who have given me help or encouragement, I am most grateful to Professor John K. Fairbank for having suggested the topic and supervised the study, to Professor Lien-sheng Yang, Harvard-Yenching Professor of Chinese History, for many invaluable suggestions, to the late David E. Owen, Gurney Professor of History, Harvard University, and to Derrick M. Wilde for having read parts of the original manuscript; and to Professor C. T. Wu of Hunter College for his advice in map-making. To the librarians, including Dr. Kaiming Chiu, former chief librarian, and the staff of the Harvard-Yenching library, and to those of the Houghton and Widener libraries of Harvard University, the East Asiatic library of Columbia University, and the library of the School of Oriental and African Studies, London University, I offer my sincere appreciation. Acknowledgment should also be made of the publications on the sixtieth anniversary of the Chinese Post Office sent from Taipei by the Ministry of Communication.

In expressing my thanks to the East Asian Research Center for publishing this work, I want to say how grateful I am to the members of the Center's editorial department, especially Miss May Wu who edited the manuscript and Miss Diana Lee who completed the preparations for publication. Mr. William Minty redrew my sketches of maps. I wish to take this opportunity to thank all who have kindly assisted me or lent me materials for the topic at various times; but I alone am responsible for any error or shortcoming that may occur.

# CONTENTS

## MAPS

## TABLES

### *Notes on the I-chan Map and Romanization of Names*

The routes for the mounted courier service of the Ch'ing Official Post have been drawn according to the prescribed itinerary along the districts that provided stations for riders and official travelers. Limited space precludes showing all the postal stations in each district, but all the postal districts in the eighteen provinces listed in the *Ta-Ch'ing hui-tien shih-li* have been located.

As far as possible, place names are romanized according to the system adopted by the Chinese Post Office as it appears in the *Postal Atlas of China* Peking: Directorate General of Posts, 1919) and the *Atlas of the Chinese Empire* (London, Shanghai: China Inland Mission, 1908). Generally speaking, for place names below the hsien level, as well as for names of people, titles of books or Chinese terms, the Wade-Giles system of spelling, with hyphens and aspirates, is used.

# Chapter I

## THE PROBLEMS OF MODERNIZING
## POSTAL COMMUNICATION

After 1860 the Ch'ing government, humiliated by its recent flight from Peking and apprehensive of the implementation of the Tientsin Treaties and the Peking Conventions (1858-1860), began to consider the means of strengthening China against further foreign aggression.[1] In response to the request of the Grand Council for suggestions, many high officials proposed political and military reforms, while a few with progressive outlooks advocated the adoption of Western technology which was believed to be the source of wealth and power of Western nations. In the decades following the 1860s the "self-strengthening" movement began to gain ground.[2] Governors-general like Tseng Kuo-fan and Tso Tsung-t'ang established arsenals and shipyards and began other modernization projects in their provinces with the approval of the throne. In Peking the T'ung-wen kuan, a college for interpreters, was established, and in the seventies students were sent abroad to study. After much hesitation the Ch'ing government adopted steamships, railroads, and the telegraph, created a new navy, and drilled new army units in Western methods. However, none of the advocates of reform even mentioned the possibility of modernizing China's postal system at this time. It was more than a quarter of a century after Sir Robert Hart (1835-1911) of the Chinese Imperial Maritime Customs[3] first suggested it to the Tsungli Yamen in 1861 that China established a national post office along Western lines.[4]

### The Modernization of Postal Services in the West

Prior to the 1840s postal communication in Western countries was generally unsatisfactory; facilities were unevenly distributed, postage rates were expensive, and speed was not always a primary consideration.[5] Thousands of witnesses testified before the House of Commons Select Committee on Postage in 1837 that rates in England were exorbitant.[6] Payment was usually collected from the recipient, and many poor folk could not afford to pay for letters. With the advance of industrialization, the expansion of trade, and greater mobility, the volume of correspondence increased, but high postage rates led to widespread evasion of payment resulting in severe loss of revenue to the treasury. The committee reported that the high postal rates were "extremely injurious to all classes," interfering with "their progress in moral and intellectual improvement," and "their physical welfare." In addition, the high rates produced a "most serious injury to commerce, and consequently, to national prosperity."[7] After much agitation, reform came dramatically in 1840 when Great Britain adopted Sir Rowland Hill's proposal for a low uniform postage rate—a penny for all of Britain.[8] The Penny Black, the first adhesive stamp ever used for prepayment of mail articles, revolutionized postal practices throughout the world and subsequently brought about the development of philatelic interest, which has gladdened the hearts of generations of collectors, young and old. Greater efficiency in communication and an increased volume of mail marked the British experiment even before it was completed. Other countries followed the British example and effected similar reforms.[9]

The modernization of postal services entailed the following changes: postal transmission came to be looked upon as a civilizing agent in the flow of knowledge and ideas as well as of news and commodities; postal administration came directly under government monopoly and management as a public service; postal practices were completely revolutionized, including the adoption of a low and uniform postage rate based on weight (instead of on the number of sheets), the prepayment of postage by means of adhesive stamps, and the employment of the best means available for speedy transmission and distribution of mail.[10]

Simultaneous advances in steam and rail transportation as well as in social philosophy in the mid-nineteenth century not only gave impetus to postal reforms but also accounted for their success to a large extent. In the West, a convenient, safe, speedy, and economical postal service had come to be recognized as indispensable to civilized life, and it behooved the government to provide such a service. Most governments lost no time in adopting the postal reforms outlined above, because once reform was under way, the service yielded a profit, and the more efficient the service, the greater the profit. The formation of the Universal Postal Union in 1874 crowned the reform efforts in the West, and the uniform practices and arrangements greatly facilitated the international exchange of mail. It was wonderful to think that a little piece of gummed paper attached to one's letter would carry it to any corner of the world at a small cost, and that one could drop a letter at any time of the day or night into a mailbox on any street (another innovation of the 1840s), and know that it would reach its destination.

## Postal Communication and Transportation in Nineteenth-Century China

The Chinese official postal system, dating back to centuries before the Christian era, was highly developed and sophisticated. It transmitted documents and government properties and provided accommodations and traveling facilities for civil and military personnel on official business. Although the system was very efficient by pre-modern standards, it had become increasingly inadequate by the late Ch'ing period. Besides, the official postal system served exclusively the needs of the government; postal service for private individuals varied with the locale. A city-dweller could enjoy the luxury of having his mail delivered or picked up by a courier at his house or his place of business and transmitted safely to any part of China at reasonable cost by commercial letter agencies, (min) hsin-chü, called hongs by Westerners. For those living in the interior or in certain rural areas, however, courier facilities were either infrequent or non-existent. Very often letters were sent by special messengers or were entrusted to muleteers, chair-bearers, travelers and the like because couriers and letter agencies tended to operate only in large towns and along routes where there was a sufficient volume of commercial correspondence to make it profitable.

Problems of communication in China were intensified by the lack of good roads and the vast distances to be covered. Roads generally followed the natural topography and converged on Peking. With the exception of the eastern and central plains, a good portion of the north, the west, and the southwest were difficult for travel. [11] In the Yellow River valley, south of the Great Wall. many roads were made of pounded earth and the usual means of conveyance were sedan chairs, springless carts, horses, and mules. Some roads were subject to floods in late summer and early autumn. Often in Shansi and Shensi cartwheels made such deep ruts that the sides of the roads rose like banks, and the movement of vehicles became exceedingly slow when traffic was heavy. [12] Traveling in Manchuria was equally difficult. A Westerner observed that good roads were few and far between: "In summer, during the rains, mud made communication almost impossible. In winter, for four months, the ground was frozen like iron, so that communication was easier. But very often riverbeds, dry save during the rains, were the only cart roads, and so rocky were they that carts travelled in companies to render mutual help on the journey. Moreover, they travelled in company for mutual protection of brigands."[13] The northern coast had some good harbors, but from November to March many were icebound. Until icebreakers were introduced toward the end of the century shipping was usually at a standstill during those months. In the Yangtze River valley and in the south, where roads were generally narrower, wheelbarrows and chairs were used for transportation and luggage was carried by porters. Except in Fukien, which was hilly, the excellent waterways afforded cheap and convenient transportation. In the west and southwest, which were mountainous for the most part, there were few good roads. In Szechwan the hardships of travel were proverbial; rapids in the upper Yangtze River and steep mountains made navigation dangerous. Transportation in Kweichow and Yunnan was not any better; remoteness from large trade centers and seaports and the scarcity of traffic accounted chiefly for the poor condition of the roads there. In the north and northwest caravans continued to use the ancient trade routes. Camels and horses were the chief means of conveyance in the desert. Occasionally dog-sleighs were used in these parts in winter and rafts made of blown-up buffalo skins sewn together were used in summer. Travel in these regions was arduous at best, and particularly hazardous during the severe winters.

There were many navigable waterways besides the Yangtze. Most of China's rivers ran from west to east; the Grand Canal was the only major waterway connecting the north and the south, from Peking to Hangchow. The canal was very popular, even though it had gradually fallen into disrepair. The average Chinese traveler preferred to travel on inland waters rather than expose himself to the perils of the high seas. Furthermore, compared to land travel, a boat journey generally meant less discomfort.

Foreign steamers had been plying the South China Sea coast for some years before 1860, and after that date they went as far north as Newchuang and some of the inland rivers. In 1872 a government-sponsored Chinese steamship company was formed. After 1884 railroads began to make their appearance in China, and by the end of the nineteenth century Peking could be put in touch with several major cities in China in a fraction of the time required by the mounted couriers. Although most improvements in transportation were limited to the coastal provinces and the Yangtze valley, and innumerable places in the interior remained inaccessible except by rather primitive means, the progress made in the last quarter of the century did make a substantial difference in travel and communication for most areas in China proper. The Ch'ing imperial government, however, ignored these changes and continued to transport its personnel and transmit its documents by the centuries-old method. The swift mounted couriers who at top speed might cover as much as 250 miles in twenty-four hours [14] could never hope to compete with railroads or steamships where these were available.

The discrepancies between the Chinese and Western postal systems were quite evident in the sixties. Postal reform in the West was at its peak in the mid-nineteenth century. Country after country looked up to Britain for the spectacular success of its penny postage reform and modelled their own postal system along the same lines. [15] In order for China to become a modern nation, capable of maintaining peace and order at home and of conducting friendly relations and fruitful commercial relations with other nations, an Englishman like Sir Robert Hart could hardly consider anything more basic and practical than the adoption of an up-to-date postal system whose benefits had been amply demonstrated in Western countries.

## The Problem of Postal Modernization in China

The Yamen ministers had been told many times about the various benefits of the Western postal system: the convenience to the public and to trade, and increased revenue for the treasury. For China, furthermore, it became a question of sovereignty to have a national postal service, since foreign powers were beginning to encroach on the right to administer the post. After Hart first proposed the modernization of the post in 1861 he approached the Yamen repeatedly (in 1876, 1885, and 1892) about establishing a post office, but thirty-six years were to elapse before the post office materialized. Meanwhile, after the war with England and France and the Taiping rebellion, the good feeling and unity prevailing in China in the early years of the T'ung-chih restoration (1861-1874) were replaced by an uneasy atmosphere and a precarious balance of power between factions at court. The decline of imperial prestige and the lack of leadership from the throne led to a considerable degree of political decentralization and moral deterioration in public life. The foreign powers dropped their cooperative policy toward China and began to exploit her weaknesses and try to obtain concessions and advantages for themselves, stripping China of her dependencies and converting them into colonies. Increased contact with the West brought rapid changes in China; various forces were at work corroding old ideas, institutions, and the social order. Postal communication, which affected both the government and the people alike, was no exception.

While the Yamen vacillated, the task of creating a modern postal system for China became increasingly difficult. Toward the end of the century there were no less than half a dozen postal systems in China, namely, (1) the Chinese Official Post comprising the I-chan (mounted courier service) and the P'u (foot courier service); (2) the dispatch bureaus (*wen-pao chü*), first organized by the Tsungli Yamen to handle correspondence with Chinese envoys abroad, and similar bureaus established by high provincial officials for their own dispatches; (3) the (*min*) *hsin-chü*, commercial letter agencies operated by Chinese merchants for the public; (4) foreign postal agencies established in treaty ports and patronized by Westerners in China; (5) local or municipal posts, operated by municipal councils and organized by Westerners in the foreign settlements of the treaty ports; (6) the Customs Post, which evolved from the Maritime Customs service for the transmission of diplomatic correspondence between Peking and Chinkiang during the winter months into a regular postal service for the general public in the treaty ports. By the time the imperial government was ready to assume full responsibility for a unified national postal system for all China, it had to contend with the vested interests of its own people as well as of foreigners and their governments.

Postal service in China between 1860 and 1896 must be viewed in the context of China's political, economic, and cultural developments. Why did the Chinese ministers take so long to initiate the necessary and worth-while post office project? What were the Chinese postal systems like and how did they operate? To what extent were traditional institutions affected by the impact of the West? What were the problems of modernizing the Chinese postal services? What role did Sir Robert Hart and other Westerners in Chinese government service play in the process of modernization? To what extent does the problem of postal reform reflect the general condition in China in the late Ch'ing period? These are some of the questions to be examined in the course of this study.

## Chapter II

## THE CH'ING OFFICIAL POST

The premodern Chinese postal system was often referred to as the *i* or *i-chan*, even though, strictly speaking, the I-chan was only a part of the official post. However, the I-chan was so important that the term was often used to denote the entire official postal system. The character *i* written with the horse radical implies "transmission by couriers on horseback," and *chan* means "station," but *i* and *i-chan* are interchangeable, and both words can be used to designate either the institution of postal transmission by mounted couriers or the post stations.

The general word for "post" is *yu* or *yu-cheng* (lit., "postal administration"). According to the earliest Chinese etymological dictionary, the *Shuo-wen chieh-tzu* (c. A.D. 100), *yu* means literally "the buildings through which correspondence is relayed throughout the country as far as the borders."[1] Confucius employed the terms *chih* and *yu* for "post"; a number of other terms were also used in his time.[2] As the frontiers expanded after the founding of the Han empire (206 B.C. - A.D. 220), the mounted courier system also extended, and the term *i* was adopted to denote the official postal system. The term *yu* on the other hand was never really supplanted; in the *Ta-Ch'ing hui-tien* all postal matters were classified under the heading of *yu-cheng*.

The Ch'ing Official Post was the end product of a system developed and perfected over two thousand years, and it bore the stamp of antiquity. At least a millennium before the Christian era, postal transmission was already well organized in China.[3] In the Chou dynasty (1122-249 B.C.) highways were lined with post houses that were used for military purposes, official missions, and diplomatic conferences. Food and lodging were provided along the highways; the former every ten li (one li is over one-third of a mile) and the latter every thirty li. Supplies for the post houses were stored at depots stationed fifty li apart.[4] Toward the end of the Chou dynasty, when the feudal states began to achieve independence from the central government, each state developed its own postal system. Diplomatic contacts were continuous—envoys and messengers traveled back and forth between the states—and summit conferences occurred no less frequently than warfare. Facilities for meals and lodging as well as means of transportation were available in every state. Two features thus distinguish the Chinese traditional postal system: the official post existed primarily for the administrative, military, and diplomatic needs of the government, and travel facilities formed an integral part of the postal service.

Speed was stressed in the Chou postal system. Confucius was said to have remarked that "the persuasive influence of virtue spreads even faster than the transmission of royal orders by relay and postal stages."[5] The efficiency of the Chou post must have been a well known fact for him to use it as an illustration. The *Tso-chuan* makes numerous references to the swiftness of postal services in the late Chou period. For instance, when one of Ch'u's satellite states, Yung, took advantage of the famine in Ch'u to rebel and stir up the barbarians to attack Ch'u, Ch'u decided to attack Yung. Although Ch'u was a much larger state it was defeated seven times. The prince of Ch'u then traveled by post-chariot (*jih*) to meet the army at Lin-p'in. Traveling much faster than usual and without attracting any notice, he was able to surprise the enemy, and subsequently conquered Yung and pacified all the barbarians.[6]

Another feature of the traditional postal system was its military character. Beginning in the T'ang dynasty (618-907), the official post was placed under the control of the Board of War. At that time there were as many as 1,630 *i-chan*, of which 260 were boat-stations and 86 supplied both horses and boats.[7] Elaborate postal regulations and laws were formulated. The practice of placing the postal system under military control was continued down to the end of the Ch'ing dynasty. During the Sung dynasty (960-1280) soldiers replaced civilians as couriers.[8] Because of the importance attached to official correspondence, no Chinese government extended the use of its postal system to the people. The I-chan, not unlike the Pony Express in America, was reserved for important affairs of state, while all other postal matters were left to the P'u. Taxes and articles commanded for imperial use were also transported through the *i-chan*. The T'ang emperor, Ming-huang (712-750), was reputed to have used the courier system to send for fresh lichees from thousands of miles away—possibly Szechwan—to satisfy the whims of his favorite, Yang Kuei-fei. As the poet Tu Mu remarked: "A single horseman in the red dust/ and the young Consort laughs,/ but no one knows if it is the/lichees which come."[9]

For centuries the postal system remained largely as it had existed in ancient times. Even the Mongols with their vast empires across Asia merely extended the system without altering its structure.

### The Organization of the Ch'ing Official Post

In line with tradition, the Ch'ing Official Post served the triple function of postal transmission, transportation, and travel service. Main post routes radiated from Peking, connecting it with all provincial capitals where resided the governors-general, governors, commissioners, and commanders-in-chief[10] (see map and table 1). Subsidiary post routes formed numerous networks around each district. Along the five main routes, at intervals of about 100 li, were post stations (*i*) with horses and other facilities,[11] and on subsidiary routes, at intervals of approximately fifteen to twenty-five li, were foot-courier stations (*p'u*).[12] In Shengking the post stations were also called *i*, but in Inner and Outer Mongolia and in Sinkiang (before it became a province in 1884) the stations were variously known as *chan, t'ang, t'ai,* or *t'ai-chan.* They were all military postal stations and were managed by the local military authorities. In the northwest there had been a few stations known as *so* that were used chiefly for the transportation of government properties, but by the mid-nineteenth century they were abolished.[13]

The Ch'ing Official Post was a colossal structure with a varying number of postal stations at different times during the dynasty. According to the Kuang-hsü edition of the *Ta-Ch'ing hui-tien,* the I-chan had 1,956 stations covering some 80,000 li of post routes (see table 2), while the P'u had from several hundred to over a thousand stations in each province. Both systems were manned by tens of thousands of people[14] (see table 3).

### Central Administration

The Ch'ing Official Post, like its predecessors, was controlled by the Board of War.[15] The board was responsible for general policy and exercised supervisory authority over postal operation. Decisions relating to the installation or abolition of certain postal stations, the increase or decrease of quotas for horses at each station, expenses allowed for horses and personnel and other such matters were all made by the Board of War, subject to the final approval of the emperor. Specifically the Remount Department (Che-chia ch'ing-li ssu) of the Board of War was responsible for the postal service.

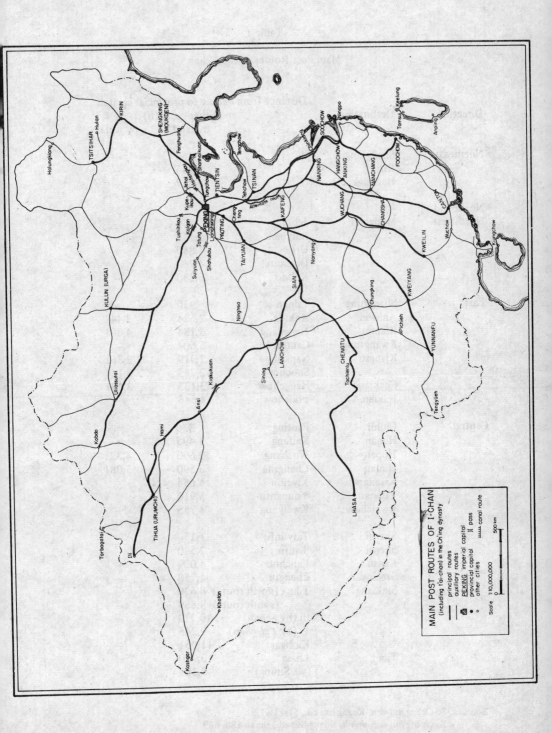

MAIN POST ROUTES OF I-CHAN
(including t'ai-chan) in the Ch'ing dynasty

principal routes
auxiliary routes
**PEKING** imperial capital
provincial capital ✕ pass
other cities ∼∼∼ canal route

Scale 1:10,000,000
0                    500 km

Table 1

Main Post Routes of the I-chan

| Direction | Destination | | Distance from Peking to provincial capitals and other cities (*li*) | |
|---|---|---|---|---|
| | | | by land | by water |
| Northeast | Shengking | | 1,460 | — |
| | Kirin | | 2,245 | — |
| | Heilungkiang | | 3,317 | — |
| North and Northwest | Mongolia | Jehol | 450 | — |
| | | Suiyuan | 1,145 | — |
| | | Urga | 2,880 | — |
| | | Uliassutai | 4,960 | — |
| | | Kobdo | 6,280 | — |
| East | Shangtung | Tsinan | 930 | —* |
| | Anhwei | Anking | 2,624 | 3,441 |
| | Kiangsi | Nanchang | 3,184 | 4,081 |
| | Kwangtung | Canton | 5,604 | — |
| | Kiangsu | Nanking | 2,319 | 2,861 |
| | | Soochow | 2,743 | 3,141 |
| | Chekiang | Hangchow | 3,133 | 3,531 |
| | Fukien | Foochow | 4,848 | — |
| Central | Chihli | Paoting | 330 | — |
| | Honan | Kaifeng | 1,495 | — |
| | Hupeh | Wuchang | 2,690 | 4,321 |
| | Hunan | Changsha | 3,590 | 5,081 |
| | Kwangsi | Kweilin | 4,654 | — |
| | Yunnan | Yunnanfu | 5,910 | — |
| | Kweichow | Kweiyang | 4,755 | — |
| West | Shansi | Taiyunfu | 1,150 | — |
| | Shensi | Sianfu | 2,540 | — |
| | Kansu | Lanchou | 4,009 | — |
| | Szechwan | Chengtu | 4,750 | — |
| | Sinkiang | Tihua (north route) | 8,639 | — |
| | | (south route) | 8,869 | — |
| | | Ili (by *i-chan*) | 10,214 | — |
| | | (by *t'ai-chan*) | 9,245 | — |
| | | Kashgar | 11,951 | — |
| | Tibet | Lhasa | 10,920 | — |
| | | (via Sining) | 8,189 | — |

Sources: *Ta-Ch'ing hui-tien*, Kuang-hsü ed., 51:11b-12.
    *Ta-Ch'ing hui-tien shih-li*, Kuang-hsü ed., chüan 688, 689.
*3,421 li given by *Ta-ch'ing hui-tien shih-li*, 688:3b.

## Table 2

### Number of Postal Stations

| Provinces and regions | Figures according to the Chia-ch'ing hui-tien | | | Figures according to the Kuang-hsu hui-tien | | |
|---|---|---|---|---|---|---|
| | I,chan | I | chan | I,chan | I | chan |
| Board of War, Central station | — | 1 | — | — | 1 | — |
| Chihli | 185 | — | — | 186 | — | — |
| Shenç'ing | — | 29 | — | — | 29 | — |
| Kirin | — | — | 38 | — | — | 38 |
| Heilungkiang | — | — | 36 | — | — | 44 |
| Shangtung | — | 139 | — | — | 139 | — |
| Shansi | 125 | — | — | 125 | — | — |
| Honan | — | 120 | — | — | 120 | — |
| Kiangsu | — | 40 | — | — | 40 | — |
| Anhwei | — | 81 | — | — | 81 | — |
| Kiangsi | — | 47 | — | — | 47 | — |
| Fukien | — | 68 | — | — | 68 | — |
| Chekiang | — | 59 | — | — | 59 | — |
| Hupeh | — | 71 | — | — | 70 | — |
| Hunan | 129 | 62 | — | 130 | 62* | — |
| Shensi | 331** | — | — | 184 | — | — |
| Kansu (including t'ang and so) | — | 65 | — | — | 65 | — |
| Szechwan | — | 10 | — | — | 10 | — |
| Kwangtung | — | 19 | — | — | 19 | — |
| Kwangsi | — | 81 | — | — | 81 | — |
| Yunnan | — | 23 | — | — | 23 | — |
| Kweichow | — | — | — | — | — | — |
| Sinkiang | — | — | — | — | 160 | — |
| Separate total | 770 | 915 | 74 | 625 | 1,074 | 82 |
| Grand total | 1,759 | | | 1,781 | | |

| | Figures according to the Chia-ch'ing hui-tien | | Figures according to the Kuang-hsu hui-tien | |
|---|---|---|---|---|
| Provinces and regions | I,chan | I chan | I,chan | I chan |
| a. Mongolian stations under adjutants of the Li-fan yüan at: | | | | |
| Hsi-feng k'ou | | 16 | | 16 |
| Ku-pei k'ou | | 10 | | 10 |
| Tu-shih k'ou | | 6 | | 6 |
| Sha-hu k'ou | | 11 | | 13 |
| b. Military postal stations in Mongolia under Military lt.-governor of the Altai military post road, (military lt.-governor of Chahar) | | | | |
| Senior deputy-governor of the marches (the military governor of Uliassutai) | | 44 | | 44 |
| Imperial agent at Urga | | 39 | | 39 |
| Imperial agent at Kobdo | | 25 | | 25 |
| | | 22 | | 22 |
| c. Military postal stations in Sinkiang under: | | | | |
| Military governor of Ili | | 12 | | — |
| Imperial agent at Tarbagatai | | 10 | | — |
| Military lt.-governor of Urumchi | | 27 | | — |
| Imperial agents at: Barkol | | 8 | | — |
| Turfan | | 8 | | — |
| Karashar | | 8 | | — |
| Kuche | | 10 | | — |
| Ush | | 3 | | — |
| Aksu | | 18 | | — |
| Yarkand | | 15 | | — |
| Khoten | | 7 | | — |
| Kashgar | | 6 | | — |
| Total | | 305 | | 175 |
| Grand total | 2,064 | | 1,956 | |

Sources: *Ta-Ch'ing hui-tien*, Chia-ch'ing ed., 39:18; Kuang-hsu ed., 51:2; *Ta-Ch'ing hui-tien shih-li*, Kuang-hsu ed., chuan 655-658.
*The Kuang-hsu edition of the *Ta-Ch'ing hui-tien* gives the figures 52, but there are 62 according to the *Ta-Ch'ing hui-tien shih-li*.
**Including *t'ang*.

## Table 3

### I-chan Quotas of Postmen, Horses, Other Animals, and Vehicles

| Provinces | Postmen | Post-horses | Oxen, mules donkeys | Carts | Post-boats |
|---|---|---|---|---|---|
| Hui-t'ung-kuan | 400 | 690 | – | 150 | – |
| Chihli | 10,470 | 7,094* | (D) 277 | 163 | – |
| Shengking | – | 980 | – | – | – |
| Kirin | – | 850 | (O) 850 | – | – |
| Heilungkiang | 10 | 692 | (O) 873 | 100 | – |
| Shantung | 7,491 | 2,554 | (D) 156 | – | – |
| Shansi | 4,565 | 3,480 | – | – | – |
| Honan | 4,798 | 3,771 | – | – | – |
| Kiangsu | 4,509 | 1,650 | – | – | 51 |
| Anhwei | 2,077 | 1,515 | – | – | 42 |
| Kiangsi | 2,904 | 768 | – | – | 26 |
| Fukien | 4,018 | – | – | – | 18 |
| Chekiang | 7,141 | 100 | – | – | 53 |
| Hupeh | 5,052 | 2,470 | – | – | 64 |
| Hunan | 3,238 | 1,443 | – | – | – |
| Shensi | 4,360 | 3,264 | – | – | – |
| Kansu | 4,272 | 6,525 | (O) 525 | 675 | – |
| Szechwan | 772 ½ | 763 | – | – | 11 |
| Kwangtung | 920 | – | – | – | 244 |
| Kwangsi | 280 | – | – | – | 46 |
| Yunnan | 1,674 | 508 | – | – | – |
| Kweichow | 2,680 | 1,160 | – | – | – |
| Total | 71,631 | 40,277 | 2,681 | 1,088 | 555 |

Source: Ta-Ch'ing hui-tien shih li, Kuang-hsu ed., chuan 690 (on attendants), 694 (on horses), 695 (on carts and boats).
*Including mules
(D) Donkeys
(O) Oxen

The Remount Department had two sub-offices; the Express Dispatch Office (Chieh-pao ch'u), which received and dispatched important documents, and the Central Office (Hui-t'ung kuan), which furnished horses for couriers from its main station, the Huang-hua I.[16] Forty mounted couriers or express riders were available for transmitting documents to the next station out of Peking.[17] Under the Remount Department there were sixteen superintendents of the post (*t'i-t'ang kuan*) who represented the provinces and supervised the transmission of ordinary documents between Peking and the provinces by foot couriers. They also supervised the compilation and printing of the *Peking Gazette* (*T'ang-pao* or *Ching-pao*), consisting of imperial edicts and other documents, and dispatch of the same to provincial authorities. Sixteen counterparts of the superintendents of the post resided in the provincial capitals.[18]

### Local Administration

Except for the main station in Peking, which was run by the Central Office of the Remount Department of the Board of War, all *i-chan* and *p'u* came directly under the control of local seal-holding magistrates—mostly district (*hsien*) magistrates, some sub-prefectural (*chou*) and department (*t'ing*) magistrates. The judicial commissioner (*an-ch'a shih*) of the province was concurrently responsible for the postal service. However, as the superior officers of the province, the governor-general and the governor were presumably accountable for its satisfactory operation. In the early part of the Ch'ing dynasty postmasters (*i-ch'eng*) were assigned to some of the stations, but the position was seldom filled in the second half of the nineteenth century.[19] Therefore, for all practical purposes, the postal stations were managed by the district magistrates.

In regions governed directly by military authorities postal stations came under military control. In Shengking the stations were administered by the Shengking Board of War under an inspector (*chien-tu*) and a deputy inspector.[20] In Kirin and Heilungkiang the stations were placed under the control of military governors. In Inner Mongolia the four adjutants (*chang-ching*) of the Court of Dependencies (Li-fan yuan) at the four passes (Hsi-feng k'ou, Tu-shih k'ou, Sha-hu k'ou, and Ku-pei k'ou) administered forty-five stations among them, while in Outer Mongolia, 130 stations were controlled by the military lieutenant-governor of Chahar (also known as the military lieutentant-governor of the Altai military post road), the military governor of Uliassutai, and the imperial agents (*ta-ch'en*) at Urga and Kobdo (see table 2).[21] The 132 military postal stations in Sinkiang which were under various military authorities came under civilian rule after Sinkiang became a province. Because Peking was so far from Sinkiang, Mongolia, and Manchuria, ordinary dispatches were carried along with the more urgent ones by mounted couriers; in such cases the ordinary documents were sent in batches.[22]

With the exception of the stations in Manchuria, *t'ai-chan* often served as places of exile for those convicted of public offence, that is, of malfeasance in office. To redeem themselves, these men paid from thirty-three to forty-four taels a month for the upkeep of the stations for a term of three years.[23] Frequently the offender was recalled before the end of the three-year term.[24] Failure to pay the fine might result in flogging and exile to even more remote areas. However, in the case of an official above the rank of a district magistrate or first captain who was genuinely poor and unable to pay, the appeal for leniency would be reviewed by the emperor himself.[25]

### The P'u

The foot-courier system was a postal service for local communication, communication between districts, and the transmission of documents to other provinces and to the imperial capital. In Ta-hsing, one of the 139 districts in Chihli, there were ten *p'u* located at: the district seat Chao-yang, Hsi-liu, Cheng-yang, Hung-men, Huang-ts'un, Hsia-ma, Pai-ts'un, Ch'ing-yün, and An-ting. Including the head postman there were thirty postmen. From the station in the district seat eastward it was 10 li to Chao-yang *p'u*, 8 li to Hsi-liu, and 18 li to Ta-huang *p'u* in T'ung-chou. From the station in the district seat southward it was 10 li to Cheng-yang *p'u*, 30 li to Hung-men *p'u*, from Hung-men *p'u* to Shih-ch'iao *p'u* of Wan-p'ing district. Also southward: 10 li to Huang-ts'un *p'u*, 15 li to T'ien-kung *p'u* in Wan-p'ing district. Southeast from Cheng-yang *p'u* it was 15 li to Hsia-ma *p'u*, 28 li to Pai-ts'un *p'u*, 22 li to Ch'ing-yün *p'u*, and 8 li to Ch'ao-ch'ü *p'u* of Wan-p'ing district.[26]

Chihli had a total of 847 *p'u* with 3,312 men employed as senior postmen (*p'u-ssu*) and postmen (*p'u-ping*). Shensi, a province somewhat on the periphery, had only 553 *p'u* and 1,885 postmen, while Hunan, which was the midpoint between north and south China and the gateway to Yunnan and Kweichow, had as many as 1,222 *p'u* and 6,544 postmen. Some districts in Hunan had as many as 150 stations, while those in Chihli never had more than ten each. [27] It is interesting to compare the ratio between the number of *i-chan* and *p'u* in each province. In the western provinces both kinds of stations were scarce, but in Chihli, where all the main roads converged, the smaller number of *p'u* was compensated for by a larger number of *i*, in fact, the largest number in all the provinces (see table 2).

Postmen or couriers for the P'u were selected from able-bodied adult males whose land tax amounted to one to two *tan* of rice a year. "They should be young and strong and must render their service personally." [28] The *Ta-Ch'ing lü-li* stipulated that there was to be one senior postman and four postmen at each station, but the number sometimes varied according to local needs.

Marco Polo's description of the Chinese couriers in the thirteenth century still fits their counterparts in the nineteenth:

> They wear girdles round their waist, to which several small bells are attached, in order that their coming may be perceived at a distance; as they run only three miles, that is from one of these foot-stations to another next adjoining, the noise serves notice of their approach, and preparation is accordingly made by fresh courier to proceed with the packet instantly upon the arrival of the former. [29]

Instead of the girdle of bells the Ch'ing courier could use a hand bell. The regulations in the *Ta-Ch'ing lü-li*, however, cautioned against using the bell at night for fear of arousing wild animals such as wolves or tigers. Every courier was to be equipped with oiled-cloth wrappers and wooden boards for keeping the documents from becoming soiled and wrinkled, a raincoat and hat, a tasselled spear and baton to protect himself, and a receipt book for the document. At each station there were to be among other things a time-piece and two registration books. Postmen were to be on duty at all times, and at night the station was to be brightly lit. When a dispatch arrived the time of arrival was registered and any abnormality of the document, if discovered at the time, was also to be noted before the document was forwarded to the next station. [30] The courier was to proceed at the average rate of 300 li per day (twenty-four hours), which meant about 12.5 li (c. 3.5 miles) per hour. The distance between stations was between ten to fifteen li or thirty li at the most. Time limits for transmission of documents between Peking and various cities in the provinces were fixed by law (see table 4). If a courier was forty-five minutes late he was given twenty strokes of the bamboo; if he was an hour and three-quarters late his penalty was doubled. The maximum punishment, however, was fifty strokes. The same penalty applied to having the cover of the document rubbed (twenty strokes) or damaged (forty strokes). Should the document be opened, concealed, or lost, the courier would be given sixty strokes with a heavy bamboo cane. If the document in question was a military dispatch, he would receive between 100 and 300 strokes. If more than ten documents were damaged within a month, the head postmen would be given forty strokes. [31]

It is difficult to tell whether the regulations were observed to the last letter in the latter part of the Ch'ing dynasty. By and large the service seemed to have run smoothly. Since only routine or local documents were transmitted through the P'u, the system did not attract as much attention from the censors and from the emperor as did the I-chan. Between 1850 and 1896, except for an edict in 1885, there had been no injunction against the P'u; it was seldom singled out for reprimand as was the I-chan. However, granted that the documents were delivered according to schedule, a glance at the timetable will suffice to show how obsolete the system had become by the latter part of the nineteenth century, especially for transmission over long distances. [32] For example, a letter from Peking to Canton sent partly by steamer via Tientsin could arrive in a fortnight, whereas sent overland via the P'u, it would take about fifty-six days.

## The I-chan

### Postal Transmission

In order to evaluate the I-chan, it is necessary to examine the postal regulations and see if the performance of the system measured up to these regulations. The term *i-chan* has been translated by Brunnert and Hagelstrom as "military post,"[33] and indeed, the operation of the I-chan was intended to be on a military basis so that official correspondence would be speedily and safely delivered. According to the *Ta-Ch'ing hui-tien* the I-chan was reserved for the transmission of important documents between Peking and the provinces and vice versa. [34] These included imperial edicts or court letters from the Grand Council sometimes containing edicts or imperial endorsements to memorials.

## The P'u System in Changsha, Hunan

| Direction from Changsha | Intervening foot-courier stations (approximately 10 li apart) | Distance (li) |
|---|---|---|
| East (to borders of Liuyang) | 1. Tung-t'un p'u† | 55 |
| | 2. Yang-lin p'u† | |
| | 3. Ting-chia p'u† | |
| | 4. Huang-hua p'u† | |
| | 5. Chang-chia p'u† | |
| North (to borders of Hsiangyin) | 6. Ch'en-chia p'u† | 90 |
| | 7. P'ing-t'ou p'u† | |
| | 8. Ch'ing-shui p'u† | |
| | 9. Ch'iao-t'ou p'u† | |
| | 10. Ku-ts'ung p'u† | |
| | 11. Jen-chia p'u† | |
| Northeast (to borders of P'ingkiang) | Ya-tz'u p'u* | |
| | Pai-mau p'u* | |
| | Ch'ih-shih p'u* | |
| | Mau-t'ang p'u* | |
| | Ch'eng-lin p'u* | |
| | Hsin-an p'u* | |
| | Ma-an p'u* | |
| | Ch'ing-shan p'u* | |
| | Sai-t'ou p'u* | |
| | Fu-lin p'u* | |
| | Ts'ung-lin p'u* | |
| | Chu-sha p'u* | |
| | Niu-chiao p'u* | |
| West (to borders of Ninghsiang) | (no fixed stations) | 118 |
| South (to border of Shanhua) | | 2 |

†The numbers correspond to those in the map.
*Stations not included in the offficial quota

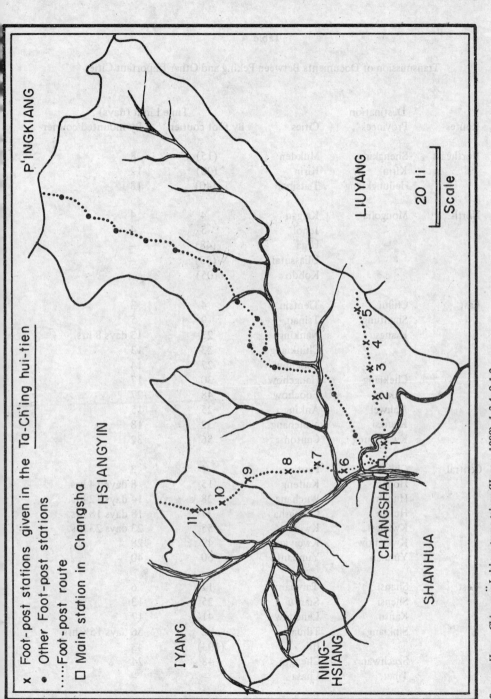

Legend:

× Foot-post stations given in the *Ta-Ch'ing hui-tien*
• Other Foot-post stations
⋯⋯ Foot-post route
□ Main station in Changsha

P'INGKIANG

LIUYANG

HSIANGYIN

IYANG

NING-HSIANG

CHANGSHA

SHANHUA

20 li
Scale

Source: *Hunan Chiang-yü I-ch'uan tsung-ts'uan* (Hunan, 1888), 1:3-4; 9:1-2.

Map of the P'u System in Changsha District

Table 4

Transmission of Documents Between Peking and Other Important Cities

| Routes | Destination Provinces | Cities | Time Limit (days) By foot courier | By mounted courier |
|--------|-----------|--------|------------------|--------------------|
| Northeast | Shengking | Mukden | (15) | 8 |
| | Kirin | Kirin | (20) | 12 |
| | Heilungkiang | Tsitsihar | (40) | 18 |
| | | | | |
| North | Mongolia | Kalgan | 4 | 4 |
| | | Jehol | 5 | 4 |
| | | Urga | (48) | — |
| | | Uliassutai | (83) | — |
| | | Kobdo | (105) | — |
| | | | | |
| East | Chihli | Tientsin | 4 | 3 |
| | Shantung | Tsinan | 9 | 5 |
| | Kiangsu | Nanking | 23 | 13 days 8 hrs. |
| | | Chinkiang | 23 | 13 |
| | | Soochow | 27 | 14 |
| | Chekiang | Hangchow | 30 | 17 |
| | Fukien | Foochow | 48 | 27 |
| | Anhwei | Anking | 25 | 15 |
| | Kiangsi | Nanchang | 32 | 18 |
| | Kwangtung | Canton | 56 | 32 |
| | | | | |
| Central | Chihli | Paoting | 4 | 3 |
| | Honan | Kaifeng | 15 | 8 days 14 hrs. |
| | Hupeh | Wuchang | 28 | 14 days 12 hrs. |
| | Hunan | Changsha | 37 | 18 days 18 hrs. |
| | Kwangsi | Kweilin | 55 | 23 days 23 hrs. |
| | Kweichow | Kweiyang | 49 | 28 |
| | Yunnan | Yunnanfu | 60 | 40 |
| | | | | |
| West | Shansi | Taiyuan | 12 | 6 |
| | Shensi | Sianfu | 25 | 13 |
| | Kansu | Lanchow | 41 | 17 |
| | Sinkiang | Tihua | 82 | 36 days 15½ hrs. |
| | | Ili | (193) | 43 |
| | Szechwan | Chengtu | 48 | 24 |
| | Tibet | Lhasa | — | — |

Sources: *Ta-Ch'ing hui-tien,* Kuang-hsu ed., 51:12b, 69:13b-15. *Ta-Ch'ing hui-tien shih-li,* Kuang-hsu ed., 700:1-4, 8b-11. John K. Fairbank and Teng Ssu-yu, "On the Transmission of Ch'ing Documents," *Harvard Journal of Asiatic Studies,* 4.1:23-35 (May 1939).

Government offices in Peking other than the Grand Council were allowed to use the I-chan only for matters of extreme urgency; ordinary documents were to be forwarded by the P'u. The use of the I-chan by provincial officials was also restricted. They were to employ their own couriers and defray their expenses. Their couriers, however, were entitled to the use of two horses at each station (usually two men were sent together for important documents) if the memorials to be presented had to do with the birthday of the emperor (or the dowager empresses), military or financial matters, recommendations for vacant posts, or impeachment. [35] In such cases the couriers were given express tallies (*huo-p'ai*) issued by the Board of War for identification and specification as to the number of horses required at the stations en route. The higher officials were allotted a certain number of tallies each year—from twenty for most governors-general down to two a year for the salt commissioner; only the judicial commissioner had unlimited use of the I-chan in connection with postal matters. The number of tallies used each year was reported to the Board of War. The board also required each province to report periodically the number of documents received from Peking and the number dispatched by the I-chan.

To ensure prompt delivery, time limits specified down to quarter hours were fixed for postal transmission between Peking and the important cities in the provinces where the higher civilian and military officials resided (see table 4). According to the degree of urgency, documents were transmitted at rates of 300 li up to 800 li per twenty-four hours. The maximum rate was seldom used, but imperial commands to "memorialize at the rate of 600 li per day" were frequently given. [36] Penalty for delay or for tampering with documents was severe; a courier might be flogged with the bamboo, dismissed, or exiled, and the official who shared the responsibility might be demoted, transferred, dismissed, or have his salary withheld. In cases involving military dispatches, the courier would be tried by martial law. [37] His superior, the district or department magistrate, would also be held responsible. [38]

Postal transmission by the I-chan was fairly satisfactory, even though the system itself gradually became more and more inadequate in the latter half of the nineteenth century. This deterioration was partly due to circumstances beyond the control of the I-chan and partly due to abuses within the system. Most complaints against the I-chan arose during the late 1850s and early 1860s when the Taiping rebellion and the temporary occupation of Peking and the exile of the court to Jehol seriously disrupted the functoning of the postal system. Even military dispatches sent at the express rate of 600 li per day were delayed from a few days up to thirty days. On one occasion in 1861 the Hsien-feng Emperor (r. 1851-1861), exasperated, asked the Board of War to have the provincial authorities investigate the delays. His edict thundered:

> The governors-general, governors, and judicial commissioners have been extremely lax in the performance of their duties, their indolence being habitual and deep-rooted. After our severe reprimands, if they are still careless and negligent, they should be brought to justice and deserve no clemency. [39]

Extensive investigations were repeatedly made. In the same year the Board of Civil Appointments was ordered to get the names of governors and judicial commissioners of Hupeh, Hunan, and Shantung, and to deliberate on the punishment to be meted out to them for failure to maintain satisfactory postal service in their provinces. [40]

When Tseng Kuo-fan (1811-1872) was encamped at Ch'i-men in southern Anhwei in 1860, he asked his brothers at Anking to send all important correspondence by one of the Hunan Braves and leave only relatively unimportant messages to be transmitted through the I-chan. [41]

Some allowance may be made for the I-chan, however, for many postal stables were burnt or destroyed, and horses were killed or captured by the Taiping rebels. [42] Moreover, re-routing the dispatches to Jehol added four days to the normal journey to Peking. Attempts were made by both the imperial and local authorities to restore postal communication as quickly as possible. For instance, in 1858 Hu Lin-yi (1812-1861), the governor of Hupeh, was allowed to replace post horses locally instead of getting replacements from official depots in the north. [43] Several p'u in Shansi and Hupeh were changed into i to expedite postal transmission. [44] Prompted by a memorial from Yang Ping-chang, a sub-expositor in the Hanlin Academy, the emperor issued an edict instructing local authorities to fill up any shortages in the fixed number of horses at each station within one month. [45] After the court returned to Peking under the T'ung-chih Emperor (r. 1862-1874), the I-chan was still far from satisfactory. Edicts were issued again and again to remind the provincial officials of previous injunctions by the late emperor on postal matters. [46] The

Board of War was asked to check each document upon arrival for the distance covered and the time taken in transit. In case of delay, the governors-general and governors were to be notified immediately so that they might investigate every station through which the document had passed and have the culprits impeached and punished. At the same time, these officials were asked not to strain the I-chan service, but to observe scrupulously the regulations in force.[47] After the mid-1860s the postal service was more or less normal again. It was not until 1894 and 1895, during the Sino-Japanese War, that Peking again complained about the delay in transmitting military dispatches.[48]

How reliable was the I-chan service? In Peking any document to be forwarded by the I-chan was first inspected and stamped by the Remount Department, which then issued an express warrant (huo-p'iao) in the name of the Board of War.[49] The document was then sent to the Express Dispatch Office where it was put under cover, sealed, placed between wooden boards, wrapped in silk, and sealed again at the seams. For greater security some documents were put in dispatch boxes, which were assigned to certain governors-general and governors, and other high provincial officials. Keys were usually sent in advance to those who were not important enough to have a regular dispatch box. The properly sealed document was then ready to be taken to the Central Station where it was forwarded to the first station along the appropriate route out of Peking for relay to its destination.[50] Documents going to Peking were accompanied by a waybill (p'ai-tan) which registered the exact time of arrival at each station and any abnormality observed at the time.[51] After they arrived at Peking, the documents were sent to the Express Dispatch Office and then forwarded to Tsou-shih ch'u (Chancery of memorials to the emperor), a division of the imperial household. "This latter examined memorials to see if they were in proper form, and if so, handed them over to the Grand Council for presentation to the emperor."[52]

After 1860 the most astonishing case of irregularity recorded was the leakage of a secret memorial presented by the Tsungli Yamen on May 14, 1874, dealing with coastal defense during the Taiwan crisis. The Yamen officials maintained that they had taken every precaution to keep the matter secret, but the text of the memorial was published in a Shanghai newspaper.[53] In the same year a memorial by Shen Pao-chen (1820-1879) also appeared in a Chinese daily in Hong Kong.[54] Even toward the end of that year, notwithstanding intensive investigation, it was still a mystery how the newspapers had obtained the texts of the memorials. Occasionally there was evidence that documents were being tampered with en route. When a case of tampering was reported to the throne, investigation was carried out in every station that had handled the document, and the culprit and his superior were duly punished.[55]

From the cases cited above, the I-chan did not seem to have always measured up to expectation, particularly in wartime, when its services were needed most. However, it would be rash to infer from these isolated instances that the system was neither safe nor speedy, just as it would be naive to assume that all was well in the absence of recorded complaints.

In their detailed study, "On the Transmission of Ch'ing Documents," Professors Fairbank and Teng remarked: "The performance of the postal service in the period 1842-60 measures up, to a surprising degree, to the standards set in the regulations in force during the whole of the nineteenth century. Documents were transmitted faster than we might have supposed."[56] They found that documents could be transmitted between Canton and Peking (5,604 li) in fifteen days, at the rate of 375 li a day, and between Nanking and Peking (2,319 li) in five days at 440 li a day (or in the record time of three days).[57] In both cases the documents arrived well within the time limits—thirty-two days for Canton and thirteen days and eight hours for Nanking by mounted couriers (see table 4). There is no reason to suppose that postal transmission via the I-chan in the decades after 1860 compared with the period 1842-1860, fell much below the prescribed standards. The Peking government continued to rely heavily on the I-chan system throughout the century. As official correspondence rose in volume, horses were added to several stations in the Tientsin area in 1875 and 1878.[58]

### Travel Service

In addition to postal transmission the I-chan also transported government funds and properties, escorted prisoners, and provided accommodation and transportation for couriers and officials on special missions. It was largely a result of the travel services that the I-chan became corrupt in the late Ch'ing period.

According to the Ta-Ch'ing hui-tien the travel services of the I-chan were restricted to the use of persons on special missions, such as examiners going to preside at the triennial examinations in the provincial capitals, generals going to war, tribute bearers, graduates of Yunnan, Kweichow, and Sinkiang coming to or returning from Peking for the metropolitan examination, widows of garrison officers who died on duty, and special messengers.[59] An identification certificate (*k'an-ho*) issued by the Board of War entitled the holder to obtain at the post-stations, according to his rank and the nature of his business, a certain number of porters, horses, vehicles, or boats.[60] The law prescribed strict punishment for those who dared to demand more than the prescribed number of men or animals or demanded animals of better quality.[61] In the case of couriers and individuals, express tallies were issued instead of indentification certificates.

Itineraries and times of arrival were predetermined by government regulations. An official usually followed the prescribed route, journeying day after day, from station to station, covering an average of 100 li per day except in hilly regions like Fukien where it was necessary to slow down. Because of the vast distances to be covered and the traveling conditions, some of the journeys took very long. In the nineteenth century it usually took fifty days to travel from Peking to Canton, but in 1861 it took Shen Kuei-fen (1817-1881), the imperial examiner, eighty-five days to make the journey because the war made a detour necessary.[62] (In normal circumstances detours were not permitted.) In 1875 it took another examiner, Ch'ü Hung-chi, who later became a grand councilor, sixteen days just to go from Peking to Kaifeng in the neighboring province; his journey from Peking to Foochow in 1891 took no less than fifty-seven days.[63] T'ao Mo, a general well known for his pacification of the northwest, traveled nearly four months in 1891 to take up his appointment as governor of Sinkiang. Prior to this journey he was commanded to leave his post in Sianfu for an audience in Peking; consequently, he spent over half of that year on the road, covering a total of 11,233 li.[64]

Most travelers began their day's journey at dawn and spent the night at post houses or inns designated for I-chan travelers in the less frequented areas. Marco Polo described the post houses he saw in Yuan China as "large and handsome buildings, having several well furnished apartments, hung with silk, and provided with everything suitable to persons of rank."[65] Six-hundred-odd years after this was written conditions were quite different, at least from the point of view of certain Chinese officials. Among the I-chan stations, some buildings were "spacious" and comfortable, and new ones were occasionally built, but many were "small, dilapidated, and dirty." Even in the metropolitan province of Chihli, many post houses, including those on the first stage from Peking, left much to be desired. [66] One examiner stayed at a post house near the Chihli-Shansi border and found the building quite big but in disrepair. The roof leaked and the tiles fell down in the heavy rain.[67]

In the absence of organized systems of public transport and the lack of amenities in inns, land travel was cumbersome. An official usually rode in a sedan-chair, accompanied by four or five servants and baggage in half a dozen carts, pack-horses, or mules; in the south and southwest, where carts were useless, baggage was carried by porters on foot. The baggage included clothes and books as well as bedding and cooking utensils. Examiners generally traveled in pairs, a chief examiner and the assistant chief examiner; thus they had twice as many servants, and double the amount of baggage.[68] Many officials complained in their diaries about the cunning of chair-bearers, cart-drivers, porters and the like who were usually hired at each station. Sometimes departures were held up because of wrangling over wages and cart rentals. One of Shen Kuei-fen's experiences on his way to Canton gives a glimpse of the nature of I-chan travel and the problems faced by both the travelers and the post stations.

When Shen and his party reached Nanyang in southern Honan, he wanted to continue his journey to the prefectural city the following day, but was advised by the magistrate's servants to proceed to Li-ho for lunch and find lodging at Wa-tien instead. He later discovered that the prefectural city was off the main route for those going south, and to stay there he had to cross and recross a river. In switching the post station from the prefectural city to Li-ho and having lunch served there, the district was not only thinking of the convenience of the travelers but was also saving expenses for itself. This was, however, not made clear to Shen at the outset.[69] Heavy rain during the night made the roads very muddy, and the next day a detour was necessary, adding twenty to thirty li to the journey. It turned out that the other road was not much better. Shen noted in his diary:

> Both men and horses were repeatedly in danger of stumbling—one
> pack-horse fell into the water, and my bedding was entirely drenched.
> After thirty li we refreshed ourselves at Li-ho; it was already after noon.

Wai-tien was still sixty li away; we could not possibly get there [by night-fall.] Since we could not stay at Li-ho, not only because the house was too small, but also because the Nien rebels appear frequently at Nanyang, and there have been alarms at night, we turned reluctantly toward the prefectural city . . . It was not easy to cross the White River as there were few boats . . . The magistrate's servants did nothing to help. We waited in the sedan-chairs by the river for two hours, and when the luggage had all been taken over, it was already dusk. The district magistrate came to call . . . I was a little out of sorts today. The magistrate also felt bad and apologized.[70]

Although Shen felt somewhat better about the incident afterwards, he wished the magistrate's servants had told him about the conditions of the roads or had prepared accommodations for him and his entourage at the prefectural city in the first place; it would have spared the tired travelers much unnecessary bother and fatigue. "But," he added, "how can one expect such thoughtfulness from these people?"[71]

## The Management of Post Stations

Shen Kuei-fen's experience was not typical, for most examiners reported that they were able to reach their destination without much trouble, in spite of the restrictions and the mode of travel. Shen's remark about "these people" points to a real weakness in the management of the post stations. The magistrate usually assigned his personal servants to supervise the post stations; one of them was appointed specifically to tend to all traveling missions. He alerted the magistrate of the approaching party, supervised the preparations for the party's reception, and provided accommodation and transportation to the next station the following morning.[72]

The magistrate's personal servants were not government employees, but their names and vitae were reported and registered at the superior yamen. Although they had no special qualifications except the ability to read and familiarity with yamen routines, they were often assigned to official jobs and helped the magistrate to supervise the clerks and runners.[73] The runners, unlike the servants who followed the magistrate from post to post, were natives of the district employed to perform menial tasks in the yamen. Some served as guards for escort duty, others as couriers, or as "swift hands," that is, grooms and foot messengers.[74] As a class the reputation of the runners was not enviable; many of them used their position in the government to line their pockets.[75] Since personal servants and yamen runners received very low pay, they lived on perquisites, the "customary fees,"[76] and many did not hesitate to manipulate the funds they handled. Some servants were known to have deducted sums from the runners' wages,[77] and the runners would regain their losses from something else. Once the clerks, runners, and personal servants collaborated in malpractice, it was very difficult for the magistrate to check them. The post stations with funds for wages and miscellaneous items such as fodder, medication for horses, wages for temporary porters, cart rentals, and so forth were practically open to corruption. Of course, there were servants whose behavior was exemplary. When Shen Kuei-fen was in Kwangtung he offered the governor's servant four silver dollars to bring him some documents (he gave the same amount to the clerk for copying them). The servant refused to take the money. When he was offered a piece of material for a gown and a small knife, he likewise refused. Finally, he accepted only a fan and a pair of scrolls with Shen's calligraphy.[78] In spite of their lowly status, personal servants often acquired polished manners through contact with officials. Many took great pride in their work; it was not unusual for some of them to serve a family for two or even three generations, and the faithful servant is a familiar figure in Chinese history and folklore. Lack of integrity on the part of the servant reflected not only on the servant's character, but also on that of his master.

Although the magistrate was personally responsible for the postal service in his district, his part in actual management was nominal. Whenever a high official traveling on a mission for the emperor or on his way to the capital happened to stop at a provincial capital the high provincial authorities from the governor-general down to the magistrate all paid their respects to the emperor in the person of the official before greeting each other.[79] All district magistrates were required by etiquette to welcome or bid farewell to officials of higher rank who arrived at or passed by their district.[80] Should the post station be too far from the district seat, the magistrate would send at least a messenger. The traveling officials often returned the calls. If a magistrate were specially attentive, he might give an elaborate dinner, or provide some other refreshment on the way, or send guards to escort the party through the district. Such meetings sometimes served to better communications between the capital and the provinces, but most encounters were purely ceremonial. On the much-traveled routes the magistrates were quite weary of such customs;[81] some thought of them as a waste of time and energy,[82] but no one seemed to have suggested any changes.

Many stations did not have the official quota of horses as prescribed by the *Ta-Ch'ing hui-tien*, and it was usual to hire horses, carts, and porters as they were needed. Sometimes in the name of corvée the common people were pressed into service as couriers and their animals or vehicles commandeered by the post stations with little or no reimbursement.[83] To curb abuses and to lighten the people's burden, some provinces organized "men and horse bureaus" maintained by contributions from the people. Some of these bureaus developed as a result of the war against the Taiping rebels,[84] but when the war was over many bureaus remained, and the extra taxes levied for their support became an unnecessary burden. In 1877 the governor-general of Szechwan, Ting Pao-chen (1820-1886), abolished most of the bureaus in his province except those on busy roads.[85] When Chang Chih-tung (1837-1909) became governor of Shansi in 1882 he reduced the corvée in more than 100 departments and districts as part of his general effort to clean up the administration[86]—something of an Augean stable! He asked the judicial commissioner to see that enough good horses were purchased by the local governments to make up the quota and to improve postal service in general. He also warned the officials that if any documents were delayed or private property violated through lack of horses, he would impeach them.[87] Chang also ordered financial reports in arrears to be cleared and monthly reports to be submitted by the districts:

How many carts and horses were used during the month and for what purposes? What was the amount commuted for traveling officials or messengers? How much was paid for horses and vehicles? To whom was this paid? How much was spent? How much was left from last month's allowance? Where is the money now?[88]

Before Chang could follow through with this program of reform, however, he was made governor-general of Kwangtung and Kwangsi early in 1884, only two years after he had been posted to Shansi.[89]

There were probably other officials who tried to clean up the administration and increase the efficiency of the postal system. However, since both the P'u and the I-chan were under the same local governments and were very much a part of local administration, any reform of the Ch'ing Official Post would have amounted to a general civil service reform involving 1,303 districts.[90] with nearly two thousand horse-stations and tens of thousands of foot-courier stations, not counting those in Mongolia and the border regions. Furthermore, the Ch'ing officials were there to administer the laws and to function within the established framework. In the late nineteenth century the political atmosphere was such that with the exception of a few high officials who had the prestige and the courage to withstand opposition and slander and who were energetic enough to initiate drastic reforms and projects, the greater part of the bureaucracy was inclined to "let sleeping dogs lie."

## Postal Expenses

While modernized postal services in the West were yielding revenue to the treasury, the Chinese Official Post constituted a sizable item in government expenditure. The I-chan alone cost the Ch'ing government nearly 2 million taels a year out of a total budget of 30 to 40 million taels (70 to 80 million taels in the last two decades of the nineteenth century). Tls. 1,959,870 were allocated to the I-chan according to the *Ta-Ch'ing hui-tien* but financial difficulties in the 1880s and 1890s necessitated some retrenchment in spending.[91] Nevertheless, in the decade 1885-1896 I-chan expenditure at its lowest was no less than Tls. 1,565,169 (1885).[92]

While I-chan expenses were especially allocated by the Board of Revenue, the expenses of the P'u and the cost of repairing and replacing post-boats came out of the local government's regular expenses.[93] The provincial officials, particularly the governors-general and governors, regularly sent two express couriers to Peking once a month. Although these couriers used the facilities of the I-chan they were paid by the provincial g vernment. Sometimes other officials also utilized their services.[94] If all other postal expenses in the empire were added to I-chan expenses the total expenditure might well be 3 million taels a year as alleged by the critics of the I-chan system.[95]

Table 5

I-chan Quotas and Expenses (in Taels)

| Province | 1818 Original quota | Misc. sums* | 1899 Original quota | Misc. sums | Actual expenses 1893 |
|---|---|---|---|---|---|
| Board of treasury | | | | | |
| Chihli | 403,401 | 7,664 | 412,464 | 7,664 | 157,218,863 |
| | — | 1,868 | — | 1,864 | 5,707.34304 |
| | — | 6,380 | — | 6,380 | — |
| | — | 955 | — | 963 | — |
| Shengking | (no specific quota) | 2,664 | (no specific quota) | 2,664 | 27,443.999 |
| Fengtien | | | | | 6,753.494 |
| Kirin | 510 | 4,365 | 510 | 4,365 | 75,607.1166 |
| Heilungkiang | 1,000 | 2,808 | 1,000 | 2,808 | 18,681.563 |
| | — | 5,000 | — | 5,000 | — |
| Shantung | 155,476 | 20,447 | 141,675 | 20,447 | 151,358.315 |
| Shansi | 105,831 | 75,645 | 105,831 | 75,645 | 191,939.426* |
| Honan | 275,768 | — | 287,075 | — | 282,388.605 |
| Kiangsu | 152,170 | — | 152,170 | — | 99,031.459 |
| Anhwei | 104,474 | — | 104,474 | — | 63,720.729 |
| Kiangsi | 109,997 | — | 109,997 | — | 69,371.400 |
| Fukien | 65,605 | — | 65,605 | — | 29,315.420 |
| Chekiang | 63,714 | — | 63,714 | — | 64,226.182 |
| Hupeh | 97,634 | — | 97,634 | — | 114,293.078 |
| Hunan | 93,489 | — | 93,489 | — | 62,227.47 |
| Shensi | 142,181 | 6,552 | 155,033 | — | 191,939.426** |
| Kansu | 43,695 | 1,305 | 47,521 | — | 144,097.367 |

Table 5 (Continued)

| | Quotas | | | | Actual |
| Province | 1818 | | 1899 | | expenses |
| | Original quota | Misc. sums* | Original quota | Misc. sums | 1893 |
| --- | --- | --- | --- | --- | --- |
| Szechwan | (no specific quota) | — | — | — | 34,784.8248 |
| Kwangtung | 5,549 | — | 5,549 | — | 10,428.447 |
| Kwangsi | 9,869 | — | 9,869 | — | 3,461.3 |
| Yunnan | (no specific quota) | — | — | — | — |
| Kweichow | (no specific quota) | — | — | — | 61,316.624 |
| Sinkiang | — | — | 106,260 | — | — |
| Totals | 1,830,363 | 136,958 | 1,959,870 | 127,800 | 1,830,905.89464 |

Source: Quotas for 1818, *Ta-Ch'ing hui-tien*, Chia-ch'ing ed., 39:20b-21b. Quotas for 1899, *Ta-Ch'ing hui-tien*, Kuang-hsu ed., 51:4-5.

For expenses, see Li Hsi-sheng, ed., *Kuang-hsu k'uai-chi-lu* (Shanghai, 1896), 3:11.

*Allowances for purchase of carts, horses, or other animals, and money commuted from rice, beans and wheat.

**There is apparently some mistake in one of these two figures given by Li Hsi-sheng. The present tabulation does not add up to the total amount of expenses indicated.

Table 6

Additional I-chan Quotas and Expenses (in Kind)

Quotas

| Province | | 1818 | | 1899 |
|---|---|---|---|---|
| Chihli | Rice | 284 *tan* | Rice | 286 *tan* |
| | Beans | 3,556 *tan* | Beans | 3,525 *tan* |
| | Hay | 21,300 bundles | Hay | 21,300 bundles |
| Kansu | Rice | 7,628 *tan* | Wheat | 543 *tan* |
| | Beans | 1,090 *tan* | | |
| | Hay | 33,120 bundles | Hay | 19,542 bundles |
| Sinkiang | | | Flour | 486,540 catties* |
| Anhwei** | Grain | 15,404 *tan* | Grain | 15,404 *tan* |
| | Wheat | 22 *tan* | Wheat | 22 *tan* |
| | Beans | 9 *tan* | Beans | 9 *tan* |

Sources: Quotas for 1818, *Ta-Ch'ing hui-tien*, Chia-ch'ing ed., 39:20b-21b. Quotas for 1899, *Ta-Ch'ing hui-tien*, Kuang-hsu ed., 51:4-5.
*Equivalent to 4,550 *tan* of wheat.
**Income from *ma-t'ien* (fields for horses)

Table 7

Annual I-chan Expenses, 1885-1894

| Year | | | Total I-chan expenditures (Tls.) |
|---|---|---|---|
| Kuang-hsü | 11 | (1885) | 1,565,169.00 |
| | 12 | (1886) | 1,576,339.28 |
| | *13 | (1887) | 1,682,455.25 |
| | 14 | (1888) | 1,579,364.00 |
| | 15 | (1889) | 1,710,956.874 |
| | *16 | (1890) | 1,849,277.860 |
| | 17 | (1891) | 1,734,709.40 |
| | *18 | (1892) | 1,819,391.40 |
| | 19 | (1893) | 1,830,905.894 |
| | 20 | (1894) | 1,759,948.41 |

Source: Liu Yüeh-yün, comp., *Kuang-hsu hui-chi-p'iao* (N.p., 1901), 1:3-4b.

*Intercalary years

*The Need for Modernization*

Notwithstanding some shortcomings and abuses, the Ch'ing Official
Post continued to render important service to the end of the century. Some
of the cases presented in the foregoing discussion might have conveyed a
one-sided impression, but it is to be remembered that an institution is more
likely to receive attention when things are going wrong rather than when all
is well. What eventually made the system obsolete in the light of advances
in postal administration in the West was the introduction of steamships and
railroads in China. But the real defects of the system were not apparent to
the Chinese officials until industrialization sufficiently changed the socio-economic
conditions in China.

A number of issues can be raised with regard to the Ch'ing Official
Post, particularly the I-chan: (1) Relating to management. Was it wise not to
have a postal officer in charge of the postal stations but pile the administrative
work onto the multifarious duties of the local magistrate and leave the
supervision to the judicial commisioners who had other responsibilities? (2)
Relating to methods of transmission. Why was no use made of steamers and
new means of transportation? Why were the roads not improved? (3) Relating
to organization. Was the travel service, which was so complicated and expensive,
really necessary? (4) Relating to aims. Should the Official Post merely serve
the government and not benefit the whole nation? Was it not desirable for
China to adopt a national postal system yielding revenue to the treasury
rather than retain a limited postal system which cost the government over 2
million taels a year?

It is quite likely that these issues were hinted at by Sir Robert Hart
when he first proposed the establishment of a modern post office for China
in 1861. Although Hart did not criticize the I-chan when he mentioned the
proposal to the Tsungli Yamen officials, the very virtue of a modern post
office showed up pointedly the need for modernization in the Chinese postal
system. However, the traditional postal system was neither modified nor
changed in the least until the end of the nineteenth century. Most Chinese
officials believed that the I-chan was by far the best postal system that ever
existed, if the regulations were only conscientiously observed.

## Chapter III

## THE LETTER AGENCIES

As late as the nineteenth century private letters were usually entrusted to friends, acquaintances, special messengers, travelers, chair-bearers, cart-drivers, or muleteers for delivery.[1] In large cities mail was delivered by regular couriers or letter shops, *(min) hsin-chü* (lit., "[people's] letter agencies").

It is not clear how the letter-carrying business originated. It is possible that individuals who acted as couriers along certain routes[2] ended up by setting up shop with assistants and letter carriers. It is possible too that commercial firms that delivered letters for their clients along with their own correspondence eventually developed the service into a sideline. According to one account the *hsin-chü* came into being in the Ming dynasty (1368-1644), when high officials were in the habit of attaching to themselves a number of personal advisers and corresponding secretaries who moved with them from post to post. Many of these secretaries were prolific writers; they corresponded regularly with their families and friends, especially those friends who occupied similar positions in other parts of the country. The first private letter agencies were said to have originated in Ningpo, the seaport of Shaohing in Chekiang, where most of these secretaries came from.[3]

### The Organization and Operation of the "Hsin-chü"

Whatever their origin might be, by the 1860s the *hsin-chü* had long since become established in various parts of China. Indeed their heyday began shortly after 1860 and continued almost until the end of the Ch'ing dynasty.[4] Although postal merchants made no conscious efforts to organize a nation-wide postal system, many networks of postal agencies sprang up around treaty ports and large cities, including those in Manchuria; some also had connections overseas. Generally speaking each agency or group of agencies operated along particular routes within specific regions—the northeast, the Yellow River valley, the Yangtze valley, the southeast, and the southwest—with cities, especially treaty ports, as distributing centers. The following description gives some idea of the division of spheres south of the Yangtze and the importance of Shanghai in particular:

The distinguishing feature of Shanghai hongs is to be found in their importance as head or central offices of almost all the large establishments which operated on the coast north of Amoy and Swatow and throughout the Yangtze Valley. In the extreme south and overlapping into the sphere which Shanghai hongs claim, the big Canton houses have the lion's share of the business; also in the west some of the Hankow hongs independent of Shanghai carry on the bulk of distributing. Some of the firms here [Shanghai] have, however, branches in Hankow and Canton, and control distirbution beyond those points; but most depend upon agencies in these places to extend their system.[5]

The spheres were not rigidly separated. For instance, while some Shanghai *hsin-chü* had branches in Hankow,[6] some Hankow firms had branches in both Shanghai and Chungking. In addition, the Hankow firms had numerous branches and agents in the interior of Hupeh and Hunan. Chungking served as an entrepot for relaying mail between the western provinces—Szechwan, Yunnan, Kweichow, Kansu—and the rest of the country. In other regions similar postal chains were formed between cities so that most important towns were provided with some sort of postal facility.

No statistics on the total number of *hsin-chü* are available; one author estimated that there were several thousand in the latter part of the Ch'ing dynasty.[7] In twenty-four treaty ports alone some three hundred agencies were registered with the Chinese government between 1896 and 1901.[8] The number of *hsin-chü* in each city varied from twelve to thirty in river ports like Chinkiang, Hankow, and Chungking, to seventy or eighty in coastal ports like Canton and Shanghai.[9] There was a good deal of specialization among agencies in each city. Some engaged in postal transmission between the ports and large cities; some confined their operation to the interior; others undertook to forward mail not only in China but also to certain places abroad.

An agency might be owned by one person or several partners. Its staff normally consisted of a manager, a bookkeeper, a cook, and one or two runners and letter carriers. Large firms sometimes employed as many as ten or fifteen men. The clerical staff often had a financial interest in the business; some were partners, others received a percentage of the earnings in lieu of or in addition to their wages.[10] On the whole their operating expenses were kept to a minimum. [11]

## Table 8

### Number of *Hsin-chü* in Treaty Ports

| Treaty Ports | 1882-1891 | 1892-1901 Registered | Unregistered | Headquarters |
|---|---|---|---|---|
| Amoy | 23 | 30 | — | |
| Canton | — | 79 | — | |
| Chefoo | 5 | 7 | 0 | Shanghai |
| Chinkiang | — | 18 | 1 | — |
| Chungking | 16 | 7 | — | Hankow (3) |
| | | | | Chungking (13) |
| Foochow | 8 | 19 | — | — |
| Hangchow | 20 | 10 | — | — |
| Hankow | 27 | 17 | — | — |
| Ichang | 3 | 6 | 0 | Hankow |
| Kiukiang | 14 | 18 | 1 | Hankow, Shanghai |
| Kiungchow | 1 | 3 | 0 | — |
| Lungchow | 0 | 0 | 0 | private, individual |
| | | | | couriers |
| Mengtsz | 0 | 0 | 0 | — |
| Nanking | — | 17 | — | — |
| Newchwang | — | 15 | 0 | — |
| Ningpo | 15(?) | 15 | — | Shanghai, Ningpo |
| Pakhoi | — | 2 | 0 | — |
| Shanghai | — | 46 | 25 | — |
| Shasi | — | 8 | — | — |
| Soochow | — | 30-40 | — | — |
| Swatow | 19 | — | — | — |
| Szemao | 0 | 0 | 0 | — |
| Tainan | — | — | — | — |
| Tamsui | — | — | — | — |
| Tientsin | — | 14 | — | — |
| Wenchow | 9 | 7 | — | — |
| Wuhu | 15 | 17 | 0 | Shanghai |
| Wuchow | — | 6 | 0 | — |
| Yatung | 0 | 0 | 0 | — |
| Yochow | — | 4 | — | Changsha |
| Known total | 175 | 395 | 27 | |

Source: *Decennial Reports, 1882-1891* (Shanghai: Inspectorate General of Customs, 1893); *Decennial Reports, 1892-1901* (Shanghai: Inspectorate General of Customs, 1906).

0 Definitely known as having none.

Table 9

Places Served by *Hsin-chü* in Chungking

| Places served by Chungking agencies | Places served by Hankow agencies in Chungking |
|---|---|
| Szechwan<br>　Chengtu and 48 principal<br>　cities and trading centers | Szechwan<br>　Chungking<br>　Fuchow<br>　Wanhsien<br>　Yunyang |
| Kweichow<br>　Kweiyang<br>　Tsunyi | 　Kweichow<br>　Wushan |
| Yunnan<br>　Laoyatan<br>　Chaotung<br>　Tungchwan | Hupeh<br>　Ichang<br>　Shasi<br>　Hankow<br>　Fancheng |
| Kansu<br>　Tsinchow (Chinchow)<br>　Lanchow | 　Laohokow<br>　Siangyang<br>　Wuchang<br>　Wusüeh |
| Shensi<br>　Yanghsien<br>　Kiangkowchen<br>　Loyang<br>　Sian<br>　Sanyuan | Hunan<br>　Tsingshih<br>　Yochow<br>　Changteh<br>　Siangtan |
| | Kiangsi<br>　Nanchang<br>　Kiukiang<br>　Hukow<br>　Chingtehchen |
| | Anhwei<br>　Anking<br>　Wuhu |
| | Chihli<br>　Peking<br>　Tientsin |
| | All principal towns in Shantung, Shansi,<br>Fukien, Kwangtung, Kwangsi and Honan. |

Source: *Decennial Reports, 1882-1891*, pp. 117-118.

Most *hsin-chü* undertook to transmit not only letters but also bank drafts, valuables, and parcels. Working either on their own or in conjunction with other agencies, some of them forwarded merchandise and baggage and supplied chairs, porters, and pack-animals for travelers.[12] A Chinese gentleman[13] told the writer that in 1908, when he was still a child, his mother died and he was sent to join his father who was then in Hunan. He was placed in care of a *hsin-chü*, and traveled from Ningpo to Changsha in the company of one of its employees on the run of the Yangtze line. Ch'i Ju-shan, a well known playwright, recalled how the letter agencies used to run errands for their customers; for example, buying a hat or a summer outfit.[14]

There was no uniform tariff; postal rates were set according to local conditions, following a graduated scale. The rates were based chiefly on the type of letter or parcel, the distance to be covered, the accessibility of the destination, the volume of postal business for that locality, and the competition among agencies for the same route. Very often considerations of accessibility outweighed distances; for the same amount letters could be carried two or three times as far along the coast or to a place easily reached by boat than they could be carried overland. Since letters were charged per cover rather than per sheet or by weight, a slight difference in weight was immaterial. Many *hsin-chü* set a fixed tariff for places they ordinarily served and left the rates for other destinations to be agreed upon or bargained for at the time of posting. Agencies in the same locality usually made similar charges for similar services and often arrived at a common tariff.

Letters were classified as ordinary letters, letters containing drafts, and letters containing valuables. Rates for ordinary letters ranged from 10 to 400 copper cash (the exchange rate of cash and silver tael or dollar varied from time to time and from region to region); it was between 1,700 and 1,500 cash per tael, 1,500 cash per tael being more common toward the end of the century.[15] Four hundred cash was about the highest fee charged by most postal establishments; it was enough to carry a letter over several hundred to a thousand miles. For letters containing drafts, coins, or valuables, the value of the contents and the risk involved were added determinants in setting the fee. For letters containing drafts the rates usually ranged from one tenth of one per cent to one per cent of the face value of the draft, or so many cash per Tls. 1,000 (see tables 7 and 8). For letters containing silver pieces or coins the

## Table 10

### Hsin-chü Rates in Kiukiang, 1891

| Destination | Ordinary letters and small parcels (cash) | Letters enclosing dollars (cash per dollar) | Letters enclosing checks (cash per Tls. 1,000) |
|---|---|---|---|
| Shanghai | 50 | 10-12 | 700 |
| Hankow | – | – | – |
| Chinkiang | 30-40 | 5-6 | 400-600 |
| Wuhu | – | – | – |
| Chian | – | – | – |
| Kanchow | – | – | – |
| Hokowchen | 100-120 | 15-20 | 1,000-1,200 |
| Iyanghsien | – | – | – |
| Kweiki | – | – | – |
| Changshuchen | – | – | – |
| Chingtehchen | – | – | – |
| Jaochow | 40-60 | 12-12 | 700-800 |
| Loping | – | – | – |
| Poyang | – | – | – |
| Nanchang | 30-40 | 10-12 | 600 |
| Wuchengchen | 20-30 | 5-6 | 600 |

Source: *Decennial Reports, 1882-1891*, p. 226.

risk involved was greater, hence the fees ranged from one half of one per cent to four per cent of the money or valuable forwarded.

The rate for parcels was about sixty cash per catty (1.33 lbs.) for delivery within the province, 200 cash for other destinations if carried by water, and 160 to 240 cash if carried overland.[16] Letters and parcels could be registered or insured for an extra fee; in case of loss, the customer was compensated. A description of the contents and a statement of their value were both written on the cover of the letter or parcel, which was usually sealed in front of the clerk of the agency. For urgent matters a special messenger could be dispatched at the client's expense.[18] To indicate express speed, one or all four corners of the envelope was sometimes burnt and feathers were stuck in the holes—signifying "post-haste at flying speed."

## Table 11

### Hsin-chü Rates in Kiukiang, 1901

| Destination | Ordinary letters and small parcels (cash) | Letters enclosing dollars (cash per dollars) | Letters enclosing checks (cash per Tls. 1,000) |
|---|---|---|---|
| Shanghai | 20-50 | 12-12 | 400 |
| Hankow | – | – | – |
| Chinkiang | 20-40 | 5-6 | 200 |
| Wuhu | – | – | – |
| Nanking | – | – | – |
| Soochow | – | – | – |
| Ningpo | – | – | – |
| Hangchow | 50-100 | 10-20 | c. 600 |
| Wenchow | – | – | – |
| Yochow | – | – | – |
| Shasi | – | – | – |
| Ichang | – | – | – |
| Chungking | – | – | – |
| Kiaochow | – | – | – |
| Chefoo | – | – | – |
| Tientsin | – | – | – |
| Newchwang | 200 | 20-40 | c. 1,000 |
| Peking | – | – | – |
| Foochow | – | – | – |
| Amoy | – | – | – |
| Swatow | – | – | – |
| Canton | – | – | – |
| Samshui | – | – | – |
| Wuchow | – | – | – |

Source: Decennial Reports, 1892-1901, I, 358-359.

By and large the hsin-chü were modest establishments, unless they handled a sizable remittance account. Chiefly owing to this last factor, their financial situation varied considerably. Generally speaking, many letter agencies enjoyed a thriving business at their main offices, but not all their branch offices did as well. Nevertheless, the branches were maintained for the over-all operation of the firm. [19] For example, in the 1880s the larger agencies in Hankow netted a profit of some Tls. 2,000 a year, [20] while branch offices of some Ningpo agencies in Swatow considered $100 (Mexican or Spanish) as a year's fair profit. [21] On the other hand, if the sixteen letter agencies in Chungking handled between them a yearly remittance of Tls. 3,000,000, as reported in 1891, [22] and if the postal rates for drafts varied from one tenth of one per cent to one per cent of their face value, each agency might make a profit ranging from Tls. 200 to Tls. 2,000.

The letter agencies used every available means of transmission, supplying special service where it was needed and using the slower channels where economy was the first object. [23] Mail was usually carried by foot couriers who covered an average of three miles an hour; [24] in special circumstances mounted couriers were employed and relays of men and horses were provided on some routes. [25] Whenever possible water transportation was used. Many agencies in large towns along the Yangtze had small post boats of their own. In Chunking the boats had a capacity of about ten picul, capable of carrying a crew of one or two and a courier and his bags. [26] In Chekiang there were paddle-boats.

After 1860, when steamers were available not only along the coast but also in inland waters, the agencies quickly utilized their services. New agencies calling themselves *lun-ch'uan hsin-chü* (steamship letter agencies) were organized.[27] Mail bags were usually entrusted to the ship's comprador for a monthly consideration, but several *hsin-chü* in one locality sometimes sent along a common courier and shared the expense. When a streamer arrived in port, carriers from various agencies would come on board to take the bags. Quite often the mail was sorted out then and there, and local letters were delivered immediately.

To avoid unfair competition and to advance their common interest, *hsin-chü* often worked together. A good example is provided by the letter agencies of Kiukiang in Kiangsi where the letter-carrying business was highly developed. [28] In 1891 there were fourteen agencies in Kiukiang, all of them branches of *hsin-chü* in Shanghai, Hankow, and other large cities. [29] Through their main offices they were in communication with every province in China by one means or another. While all fourteen agencies desired regular connection with most of the important cities in Kiangsi, they did not maintain contact with all of them individually. Instead, each agency had its special route and connections. The fourteen arranged a schedule, whereby on fixed dates each agency took its turn at dispatching couriers to carry the mailbags for all of them along a certain route. Thus maximum efficiency was achieved with minimum cost. During the tea season, one of the agencies would provide a special courier service between Kiukiang and I-ning-chou, a tea center, with scheduled dates of departure and arrival. [30]

When there was sufficient demand for regular service in an area, an agency in the neighborhood would appoint a representative there; when the volume of business warranted a branch office, one would be established. In this manner postal agencies appeared in many of Kiangsi's large villages. An agency would refuse to handle mail destined for a town within the province that was covered by another agency, but all would accept letters for any destination outside the province. Such letters were carried to the terminus of the agency's line where another agency would take over and stamp the letters with the time of arrival with a wooden chop. Postage was either prepaid in part or collected in full by the delivering agency. [31]

Similar cooperation existed to some extent among *hsin-chü* in other parts of the country, exchanging letters and forwarding them for each other at fixed rates. [32] In Wenchow and Swatow some agencies even worked collectively, charging the same rates and dividing the profit pro rata at the end of each year. [33]

## *The Popularity of the* Hsin-chü

The Chinese public thought highly of the services rendered by the *hsin-chü*, which operated above all with a flexibility that fitted in well with popular customs and business usages. For instance, the agencies kept flexible hours, frequently remaining open until midnight. They usually sent couriers to business houses to collect their correspondence. Since many firms preferred to write their letters in the evening, when the day's business was over, collections were made as close to the time of dispatch as possible for the convenience of the customers.

The *hsin-chü* method of payment also had wide appeal. It was customary for the sender and the recipient each to pay half of the postage, although in some places the entire fee was payable on delivery, and the half payment in advance was waived. Regular patrons often had a passbook in which the number of letters sent each time was entered and stamped by the agency. Accounts could be settled monthly or at the time of the three major festivals of the year—New Year, the Fifth day of the Fifth Moon, and the Fifteenth

Day of the Eighth Moon. Some business firms found it more convenient to arrange with an agency for the conveyance of their mail at a fixed sum per annum. It was not unusual that the relationship between the *hsin-chü* and their clients rested more or less on a personal basis. In the circumstances, the regular postage rates were reduced for favorite customers, and personal letters of the employees of business firms or important customers were carried free of charge. In short, the agencies tried their best to meet the requirements of their clients and, in face of keen competition, offered all kinds of inducements to retain their patronage.

In the opinion of all who had ever used their service, one outstanding virtue of the *hsin-chü* was their reliability. There was no question but that mail entrusted to them was safe. "It rarely happened that any matter of value goes astray through the inattention or dishonesty of the couriers or their principals," wrote Commissioner Cecil Bowra of the Chinese Maritime Customs.[34] Even in circumstances that involved a certain amount of danger, such as going through the Yangtze rapids, accidents were said to be rare. Another commissioner in Ichang reported in 1891 as follows:

> As far as can be ascertained, letters are never lost except through an accident to a boat, and that occurs very rarely indeed. For instance the Ho-ch'ang postal agency is said to have employed 4,000 post-boats during the last 15 years, and only three boats with mail have, it is said, been lost during the period.[35]

Great care was taken to safeguard articles of mail. When transported by water, they were wrapped in oiled paper and packed in waterproof materials or in tins. In conveying large amounts of silver and other valuables, the *hsin-chü* often worked with the escort-agencies (*piao-chü*) that provided armed escort or convoy against robbery.[36]

As the century advanced, the reputation of the *hsin-chü* was greatly enhanced by the improvements they made in their services. C. Lenox Simpson, Commissioner of Customs at Amoy, summed up popular opinion in the *Decennial Reports* of 1901 as follows:

> The excellent manner in which these establishments are carried on has become proverbial to all of those who have come into contact with them and their methods. Letters may be sent to any points of China, through any one of them, with perfect security, and gold and silver, as well as drafts and other valuable documents, entrusted to them

under insurance. In case of loss—a rare occurrence—payment is made
under circumstances in which it might often be evaded. [37]

### New Developments After 1860

#### Remittance Banks

The period after 1860, especially the last quarter of the nineteenth century, witnessed the beginnings of industrialization and modernization in China
as well as the growth of foreign trade and increasing contact with the West. The
*hsin-chü* seized every opportunity to increase their business and adapt their
operations to changing conditions, especially in the area of their relations with
the native banks, the vernacular press, and overseas Chinese.

The old-style remittance banks (*p'iao-chuang*, lit., "draft banks") were
among the most important commercial firms which patronized the *hsin-chü*.
These banks, which had existed before 1800, began to flourish in the late Ch'ing
period. [38]  Many of them had as much capital as hundreds of thousands of
taels. Some of them became fiscal agents of the national and provincial treasuries.
By the 1880s they were found in most important cities in China, and some
extended their operations to Moscow, Kobe, Osaka, and places in Southeast
Asia. [39]  Since their main business was the transfer of funds, the banks engaged in a considerable amount of correspondence. Whenever a sum of money
was transferred by draft or bank order, by letter of credit or letter of advice,
the remitting bank sent written notice for payment to its branch or agent
where the money was being transmitted. The *hsin-chü* delivered practically
all bank mail, including letters containing silver. [40]  The close relations between the banks and the *hsin-chü* can be demonstrated by the fact that when
a *hsin-chü* was temporarily short of funds and could not meet the needs of its
own remittance service, the banks would frequently come to its assistance.[41]

After 1860 more and more Chinese moved to Southeast Asia, Australia,
and the Americas to engage in trade or to work on plantations, mines, or railroads. Since China did not have a national postal system for public use, post
offices in foreign countries could not forward any mail to the interior or to villages in China without sending it through an intermediary in the treaty ports,
and the Chinese emigrants were usually not conversant enough with the
foreign language to address envelopes or fill out forms for remittance. They
therefore sent letters and money home to their families through their "headman"—

a clan leader or someone who had brought them overseas. The headman packed all the letters in a bundle and sent them through the foreign post office to a letter merchant or *hsin-chü* at Amoy or Swatow, since the majority of overseas Chinese were natives of Fukien or Kwangtung. The agent or agency would then distribute the letters, many of them containing money, to recipients in the surrounding villages. [42] Postal fees varied according to the locality; for letters containing money postage ranged from approximately 2 to 3 per cent to 3 to 4 per cent or more of the value of the money remitted. [43]

### Gazettes and Newspapers

The *Peking Gazette* (*Ch'ing-pao* or *T'ang-pao*) was compiled by the superintendents of posts who represented the provincial governments in the Board of War. The gazette contained selections from the court circular, *Kung-mun-ch'ao*, on court activities, appointments, audiences, as well as edicts, rescripts, and memorials, and was dispatched by relays via the P'u to various yamen in the provinces. [44] With tacit official sanction some commercial firms made reprints of the gazette or compiled similar versions, and sold them in and out of the capital at great profit. [45] The *hsin-chü* helped to distribute the reprints throughout the country, and several agencies were established in Liang-hsiang, some sixty li southwest of Peking, to speed up the relays. The reprints, known as *Liang-hsiang pao*, often reached their destinations before the official *Peking Gazette* did, and were often bought by the officials themselves. In Peking a monthly subscription cost only 200 cash, but in the provinces a *hsin-chü* would ask as much as 2,000 to 5,000 cash a month. [46]

After the first Chinese daily newspaper, *Chung-wai hsin-pao* (China and foreign news daily), was published in Hong Kong in 1858, [47] a number of other Chinese dailies were published in Hong Kong, Canton, Shanghai, and Hankow. [48] When these papers first appeared, there was some difficulty selling them, [49] and the *hsin-chü* were induced to help with the distribution. *Shen pao*, one of the earliest Chinese newspapers in Shanghai, offered a 25 per cent commission to any one who would buy one or two hundred copies per day with the unsold copies returnable free of charge; the paper was sold for eight cash per copy locally and for ten out of town.[50] Many *hsin-chü* not only undertook to transmit newspapers from one city to another but eventually acted as news agents, selling papers and collecting subscriptions. [51] The profit might seem rather small at the beginning, but the Sino-French War (1884-1885) gave vernacular newspapers a tremendous boost, and thereafter sales increased significantly and steadily. [52] It was reported in Chinkiang that "what in 1881 was the exception became a rule in 1891, that all good families in Chinkiang as well as in the interior, that is, for every intelligent adult to take a glance at the Chinese daily newspaper brought from Shanghai." [53]

Although transportation in China was far from modernized and travel in many parts of the country was still very difficult, the introduction of steamships and railroads greatly facilitated communications between a number of large cities and many treaty ports where the *hsin-chü* received the bulk of their business. Together with some of the new business opportunities, the improvement of transportation enabled the *hsin-chü* to operate with increasing efficiency. They were able to deliver their mail with greater speed, regularity, and frequency and at much lower cost toward the end of the nineteenth century.

### An Appraisal of the Hsin-chü

Was the Chinese public well served by the *hsin-chü*? Were national interests well served by a commercial postal service? Was there any need for a national postal system as in Western countries? For as much as it was demonstrated that the Chinese public was quite satisfied with services of the *hsin-chü*, there were also many drawbacks to the system.

In spite of the excellent service rendered by the *hsin-chü*, they were essentially private commercial firms aiming at profit rather than public service. They tended to operate only in large towns where they could expect a profitable business. Many places in the interior had no postal facilities, not even a courier. [54] Moreover, notwithstanding the great progress they made, in places other than those frequently served or easily accessible, the postage rates were still high and delivery was slow. Finally, although some agencies in Amoy and Swatow had connections overseas, in the absence of a national postal system it was not feasible to arrange any international exchange of mail.

China definitely needed a national postal service along Western lines in order that every one in the nation could enjoy the fruits of civilization through proper communication. Paradoxically, however, the existence of an efficient commercial service such as the *hsin-chü* blinded the Ch'ing officials to the necessity of postal modernization and postponed the establishment of a modern post office for over three decades. While the officials procrastinated, the *hsin-chü* developed and prospered, and their vested interests presented great obstacles to progress.

## Chapter IV

## FOREIGN POSTAL ESTABLISHMENTS IN CHINA (1834-1896)

Today a postal service usually implies a governmental institution, but in nineteenth-century China, an anomaly existed whereby any foreign government, community, or individual could establish a postal service. The Ch'ing government was not aware of the fact that foreign postal establishments on Chinese soil encroached on its sovereignty and deprived it of a source of revenue. Since it assumed no responsibility for the postal communication of the Chinese people, it felt no more obliged to provide postal facilities for foreigners in China. Consequently, to meet their own needs, Westerners in China began to organize postal services of some kind for themselves. The British were the first and were soon followed by other nationalities. Gradually more and more foreign postal agencies were established in China for reasons of convenience, profit, or national prestige, but until their unbridled growth seriously threatened the modernization of the Chinese postal system, the Ch'ing government took no notice of them whatsoever.

### Lack of Postal Facilities for Foreign Mail

At the beginning of the nineteenth century foreign trade was confined to Canton and foreign residence to Macao and, during the trading season, to Canton. In 1836 there were 307 adult male foreign residents in Canton only 213 of them were non-Asiatic, representing forty-four Western firms. [1] Postal facilities were almost non-existent. Foreigners had to rely on ships of their own company or those of other firms for mail. However, if a letter arrived in China by any ship other than one belonging to the addressee's own firm, it was not to be delivered until after the ship's departure, "lest news of commercial value should be brought to the detriment of the firm to which the ship was consigned." [2] Sometimes a letter was delayed for a month or two or even three months. As for outgoing mail for other parts of the world, most ships leaving China were willing to carry letters for any one who wished to avail himself of their service, [3] but the letters might be subject to delay on the other end in the same manner as the incoming mail.

Between 1818 and 1833 an average of fifty-six British and thirty-seven American ships visited China yearly; the number of ships of other nationalities was negligible.[4] Since the British held the largest share of the China trade, they took the initiative in organizing the postal establishments. When the monopoly of the East India Company ended in 1834, Lord Napier arrived at Canton as the chief superintendent of British trade. On his advice the British Chamber of Commerce in Canton undertook to improve its postal arrangements and established some kind of post office in Canton. A few years later it purchased a number of boats for transmission of mail between Canton and Macao.[5]

During the Opium War in 1840 the Earl of Auckland, governor-general of India, made arrangements for ships sailing between India and China to convey mail for the British troops, and at the same time appointed a postmaster in charge of the mail at Macao.[6] When the British formally occupied Hong Kong on January 26, 1841, Captain Elliot proceeded to organize a government for the island,[7] but apparently no post office was established until Sir Henry Pottinger arrived in August to replace Elliot as the sole British plenipotentiary, minister-extraordinary, and chief superintendent of trade. [8] The Hong Kong Post Office was officially opened to the public on April 15, 1842.[9] For a while, until transmission by steamer was organized, the service was completely free for both letters and parcels. [10] When Hong Kong became a British Crown Colony in April 1843, this post office came under the jurisdiction of the British post-master general in London. [11] The informal atmosphere of the Hong Kong Post Office in its early years was described by the noted Sinologue, the Reverend James Legge:

> I used to walk to the Post Office whenever there was any arrival in the harbour by which I might expect a letter. If there were any letters for me I got them; and then the postmaster would say, "Here are letters also for so-and-so, and so-and-so, and so-and-so in your neighborhood. Please oblige me by taking them with you, and sending your coolie on with them."[12]

## British Packet-Agencies and the P. & O. Mail Service, 1842-1860

During 1843 and 1844, in accordance with the Nanking Treaty of 1842, British consuls took up residence in the five treaty ports newly opened to foreign trade and residence: Canton, Shanghai, Foochow, Ningpo, and Amoy.

To facilitate the transmission of official dispatches and general communication between Europe and China, the British postmaster-general, then Lord Lowther, suggested that the consuls act as mail agents. [13] By paying the required postage anyone could have his mail sent to the consulate, which undertook to transmit letters to British consulates in other treaty ports or to Hong Kong, and from there to be forwarded to Europe or elsewhere. Similarly, British postal agencies were established in Shanghai, Ningpo, Foochow, and Amoy. Including the agency previously established in Canton in 1834, Britain now had five postal establishments in China. Since mail was transported by packet steamers, these establishments were called packet-agencies.

The British authorities paid considerable attention to mail matters for dispatches from Canton to London had taken four to six months via the Cape of Good Hope before the shorter overland route via the Suez Isthmus was adopted in 1835. [14] Even then mail to China by a combination of steamer and clipper ship took two to three months. [15] In 1844 the average time taken for thirteen mails from London to reach Hong Kong was eighty-four days. [16] Since there was no direct mail service from England to China, the mail was at the mercy of sailing schedules, and delays were often caused by poor connections or stopovers in India and ports in Southeast Asia. After 1842, with new opportunities for commercial expansion offered by the Nanking Treaty, it was imperative to speed up postal communications.

In 1845, with a subsidy from the British government, the Peninsular and Oriental Steam Navigation Company inaugurated a monthly steamer mail service directly from Southampton to Hong Kong. [17] Since mail from London reached Southampton in a few hours, this meant direct service from London to Hong Kong. The mail traveled over water and land—by steamer from Southampton to Alexandria, by van (later by rail) across Egypt, by steamer again from Suez to Ceylon, and from Ceylon to Hong Kong.

A new era began with the arrival in Hong Kong of the wooden paddle-steamer, the *Lady Mary Wood*, on August 13, 1845, carrying the first direct mail from London in just fifty-five days. [18] The service was so successful that the British communities persistently petitioned the company to extend it to Canton and Shanghai. It was pointed out that monsoons often delayed the mail for as much as three weeks after its arrival in Hong Kong and such uncertainty was injurious to trade. [19] Finally, in August 1850, faster steamer service was extended to Shanghai by the P. & O. Co. which also sent smaller vessels to Macao and Canton. [20] In 1853 P. & O. steamers were sailing from Southampton fortnightly, and the time required for the trip was greatly shortened. [21] In 1860 mail from London was reaching Hong Kong in forty-three to forty-six days and Shanghai in approximately fifty days. [22]

The British authorities also maintained a monthly overland courier service between Canton and the British consulates in Amoy and Foochow. A round trip between Canton and Foochow took thirty-two to thirty-four days overland, but only a few days by sea. High costs prevented extension of the service to Shanghai. Since transportation by water was by far the cheapest method of postal transmission, the overland service was discontinued some time after 1850. [23]

*Foreign Mail Lines and Foreign Post Offices, 1860-1896*

By the terms of the Tientsin treaties and the Peking Convention (1858-1860) China agreed to open eleven more ports to trade; these were situated not just along the coast, but midway up the Yangtze—Newchuang, Tengchow (later replaced by Chefoo), Hankow, Kiukiang, Chinkiang, Taiwanfu, Tamsui, Swatow, Kiungchow, Nanking, and Tientsin. The number of Westerners in China in 1859 was nearly double the figure for 1855, [24] and after peace was restored in 1860 traders and missionaries came in ever increasing numbers. A sixth British packet-agency was opened in 1861 in Swatow. [25] In the same year the volume of mail of the British agency in Shanghai warranted its separation from the British consulate, and the packet-agency became a full-fledged post office. The British postal authorities in London who were in control of all of the agencies made a survey in 1867 and reported that the agency in Shanghai was the only one of the six to do any considerable business; consequently no more agencies were opened for some time. [26] In fact, the British opened only three more in the next thirty years—at Hankow in 1872, Kiungchow in 1873, and Tientsin in 1882. The 1867 survey also led the British postmaster general to relinquish control of the packet-agencies in China and place them under the Hong Kong postmaster general in 1868. [27]

The Hong Kong postal authorities were quite active even before they assumed control of all the British packet-agencies in China. In 1861 they arranged for British and French steamers to alternate in conveying the mail from Shanghai to the northern ports. [28] They also tried to secure for the British-subsidized P. & O. mail boats the monopoly of the run from Hong Kong to Shanghai. Ad- ·

hering to the concept of a state monopoly of postal services, they also tried to eliminate what they considered the "illegal" or "irregular" conveyance of letters to and from Hong Kong by private steamers. [29] However, British merchants in Shanghai continued to send letters outside of the "regular mail," since ships of various companies left daily for Hong Kong and often made better connections than the fortnightly P. & O. mail steamers. [30]

After 1860 the British postal establishments began to encounter competition from other nations when several steamship lines with government financial supports extended mail services to China. The inauguration of a French steamer service to China by the Compagnie Messageries Impériales in 1862 (renamed the Messageries Maritimes in 1871) [31] was followed in 1867 by the introduction of the Pacific Mail Steamship Company's monthly service from San Francisco to Hong Kong. The first trip of the Pacific Mail Line's steamer *Colorado* took thirty days eleven hours and ten minutes, including detention at Yokohama. [32] This line was soon in keen competition with mail lines from Southampton and Marseilles plying the Suez route, offering a cheaper and faster service. [33] In 1872 the United States Congress, which had sanctioned the subsidy for the Pacific Mail Line, became convinced of the growing political and commercial importance of communications between the United States and China and authorized a fortnightly instead of monthly service of the line to the Far East. [34] In 1875 the Japanese Mitsubishi Steamship Company bought the Yokohama-Shanghai branch of the Pacific Mail, [35] and in 1885 Germany, a late-comer in overseas expansion, also had direct steamer mail service to China through the Nordeutscher Lloyd Company. [36]

As each of these foreign shipping lines was extended to Chinese waters, a postal agency was established under the respective foreign consulate in Shanghai, the terminus of various lines and the most important of all the treaty ports. [37] A French postal agency was opened in Shanghai in 1862, an American agency in 1867, a Japanese agency in 1876, and a German agency in 1886 (see table 11). Japan also opened eight branch offices in other ports, but for lack of business all eight were closed down in 1881. For a brief period Japan, Great Britain, France, and Germany each had a postal agency in Tientsin, but the Japanese and British agencies were discontinued in 1881 and 1890 respectively. [38]

**Table 12**

Foreign Postal Establishments in China up to 1896 with Dates of Establishment

| Location | British | Russian | French | American | Japanese | German | Local or private |
|---|---|---|---|---|---|---|---|
| Canton | 1834 | – | – | – | – | – | – |
| Shanghai | 1843 | – | 1862 | 1867 | 1876 | 1886 | 1864 |
| Ningpo | 1843 | – | 1865 (agent) | | 1876* | – | – |
| Foochow | 1844 | – | – | – | 1876* | – | 1895 |
| Amoy | 1844 | – | – | – | 1896 | – | 1895 |
| Swatow | 1861 | – | – | – | – | – | – |
| Chefoo | – | – | – | – | 1876* | 1892 | 1894 |
| Chinkiang | – | – | – | – | 1876* | – | 1894 |
| Tientsin | 1882-1890** | – | 1889 | – | 1876* | 1889 | – |
| Kiukiang | – | – | – | – | 1876* | – | 1894 |
| Hankow | 1872 | – | – | – | 1876* | – | 1877 |
| Kiungchow | 1873 | – | – | – | – | – | – |
| Newchwang | – | – | – | – | 1876* | – | – |
| Wuhu | – | – | – | – | – | – | 1894 |
| Ichang | – | – | – | – | – | – | 1894 |
| Chungking | – | – | – | – | 1896 | – | 1893 |
| Soochow | – | – | – | – | 1896 | – | – |
| Hangchow | – | – | – | – | 1896 | – | – |
| Shasi | – | – | – | – | 1896 | – | – |
| Nanking | – | – | – | – | – | – | 1896 |
| | | | | | | | |
| Peking | – | 1859(?) | – | – | – | – | – |
| Whampoa | 1858(9)-1863 | – | – | – | – | – | – |

*These branch offices were closed in 1881 and, with the exception of the office in Ningpo, reopened between 1896 and 1905.

**Reopened in 1906 under the management of the China Engineering and Mining Company.

Source: G.T. Bishop, C.S. Morton, and W. Sayers, *Hongkong and the Treaty Ports: Postal History and Postal Markings*, rev. H.E. Lobdell and A.E. Hopkins (London: The Postal History Society, 1949), appendices, pp. 165-174.

The treaty ports are arranged more or less according to the date of their actual opening. See Stanley F. Wright, *Hart and the Chinese Customs* (Belfast: William Mullen & Son, for Queen's University, Belfast, 1950), appendix 3, pp. 894-895.

*Local Posts in Foreign Settlements*

As Chinese cities were opened for trade, a certain area in each city was set aside for foreign residence and business purposes. These areas, known as settlements or concessions—such as the International Settlement (originally British) and the French Concession in Shanghai, were held singly or jointly by one or more foreign powers. [39] Such areas were almost entirely free from Chinese jurisdiction, and the foreign residents in the settlements enjoyed extraterritoriality. They also enjoyed the privilege of self-government, organizing a municipal council, and maintaining their own police, fire, sanitation, and public works departments. Chinese living in the settlements also came under the control of the municipal council. Since the Chinese government provided no public postal service, some of the municipal councils organized their own postal services and even issued stamps.

The Shanghai Municipal Council was the first among foreign communities in China to organize a postal service, the Shanghai Local Post, which opened in 1864. The following year stamps were issued, but the subscription system, which contracted the Local Post to carry mail for a firm or an individual at a fixed sum per annum, was also in force. The Hong Kong Post Office sought an agreement with the Shanghai Local Post whereby the latter would deal with all local and neighborhood mail but leave all mail for southern ports to the British packet-agency in Shanghai. [40] The agreement, however, did not seem to have worked for very long. The Local Post took advantage of free transport of mailbags by foreign coastal or river steamers and soon established branches in Ningpo, Foochow, Amoy, Chefoo, Hankow, and other Yangtze ports. By 1892 the Shanghai Local Post had become a thriving institution, so prosperous and efficient that it even entertained the idea of joining the Universal Postal Union, although it was not a national post. [41]

Encouraged by its progress and prosperity, the Shanghai Local Post discontinued the subscription system in January 1893 and issued stamps that were compulsory not only for local letters, but also for letters sent through the Local Post for delivery in other treaty ports. [42] The Shanghai Local Post printed over its own stamps the name of the port to which the letter was going and put on a surcharge, often twice the value of the stamp. This provoked much opposition from its customers and from municipal councils in foreign settlements of other treaty ports. Some of these councils retaliated by taking over the branches of the Shanghai Local Post in their territories; others organized their own posts and issued their own stamps. Up to this time it seemed that the foreign settlement in Hankow was the only one besides the one in Shanghai to have a local post, but as a result of the aggressive move of the Shanghai Local Post, other local posts sprang up in Chefoo, Chungking, Chinkiang, Wuhu, and Ichang between 1893 and 1894.[43] Others probably sprang up around the same time in Foochow, Amoy, Nanking and Kiukiang (see table 11).

Many municipal councils in foreign settlements soon discovered that their local posts produced handsome revenue for them through the normal postal business as well as through the sale of stamps. Adhesive stamps, which came into use for the first time in 1840 in England as a means of prepayment, had become collectors' items. For instance, when the municipal council of Kiukiang issued 100,000 half-cent stamps for its local post in 1894, applications for 150,000 were received (one speculator wanted 95,000). When a second lot was issued, it also sold rapidly. Enormous profits were reaped by stamp dealers, private buyers, and speculators. In less than three months after the first issue was printed, the half-cent stamps were being offered for sale at ten cents a apiece.[44]

Stamps of this type are known to stamp-collectors as the "Treaty Port Local." Some municipal councils were not above exploiting this easy source of revenue by bringing out new issues frequently. The frequency of new issues soon brought ridicule and severe stricture for the local posts from many philatelic journals and collectors. However, no criticism could deter collectors from continuing to buy new stamps or speculators from exploiting this newly discovered source of wealth.[45]

With steamers of all nations calling regularly at major ports, trans-oceanic mail lines developed fast and direct postal communication between Western countries and the treaty ports and big towns in China. In the interior, however, private couriers and native *hsin-chü* were still used to transport correspondence and parcels to and from the nearest river or seaport.[46] Therefore, in spite of the large number of foreign postal establishments set up in China during the nineteenth century, the problem of foreign mail had yet to be solved.

### The Peking-Kiakhta Overland Service

While the Chinese government appeared to ignore what was happening in the foreign settlements of the treaty ports, it was apprehensive of any foreign power establishing a postal agency in the interior. The Tientsin Treaty with Russia, concluded on June 13, 1858, established a regular monthly service[47] between Peking and Kiakhta for the transmission of correspondence between the Chinese and Russian governments and for the convenience of the Russian Ecclesiastical Mission in Peking, which was established by the Treaty of Kiakhta in 1772.[48] In addition, a parcel post service was to operate four times a year between the two cities. Transmission of mail and parcels was to be undertaken by the *t'ai-chan*, but the two governments were to share expenses.

Article 12 of the Additional Treaty of Peking negotiated with Russia in 1860 reaffirmed and further elaborated the original agreement. The Chinese officials wanted to make the terms of the overland service more specific.[49] The Chinese text contained the following clause concerning the postal communications of Russian merchants trading at the frontier:

> If the Russian merchants, in sending letters, articles, and boxes for their business affairs, wish to hire men themselves and make separate regulations they shall be allowed to do so after having fully reported in advance to the local authorities and having obtained their permissions thus saving the government from paying the expenses.[50]

The Tientsin Treaty stipulated that of its three texts—Manchu, Chinese, and Russian—the Manchu text was to be regarded as authoritative,[51] but the Additional Treaty of Peking signed in Russian and Chinese did not stipulate which of the two versions was authoritative. Considerable difficulty subsequently arose from the ambiguous nature of the article just quoted.

In 1861 Russian officials at Kiakhta notified Chinese frontier authorities that twenty Russian merchants were on their way to China and requested that their application for various facilities and specifically for permission "to establish postal stations" be granted. To the Chinese officials establishing postal stations seemed totally different from "hiring men" at their own expense to transmit letters and articles; they feared for one thing that permission for Russians to establish postal stations would lead to the erection of buildings and interference with the preservation of pastures and the nomadic life of the

Mongolians. Furthermore, the Russian merchants had requested that a Russian merchant be permitted to reside at Urga and that trading be carried on at Peking instead of at the frontier. The Chinese imperial agent (*K'u-lun pan-shih ta-ch'en*) together with the assistant agent at Urga memorialized that these requests were in conflict with the terms of the treaty, and suggested that the Tsungli Yamen discuss the requests with the Russian envoy in Peking, Major Ignatiev. [52]

Debates on the interpretation of the treaty took place between Chinese and Russian officials both at the frontier and in Peking.[53] Each side argued from the text in its own language; meanwhile, the party of 20 merchants, 20 priests, 215 camels, 7 horses, and 13 carts was on its way to Kalgan with the intention of proceeding to Peking.[54] While the discussion dragged on, the Tsungli Yamen, headed by Prince Kung, began to question whether the Russian demands were inspired by misconstruction of the treaty or whether parts of the treaty had been actually and deliberately mistranslated by the Russian interpreter during the treaty negotiations.[55] The Yamen had no idea that the Russian text contained the word *posta*, which could be taken to mean "the establishment of a postal service."[56] The Chinese government was adamant in its refusal to grant the requests of the Russian merchants, who were eventually persuaded to trade in Tientsin with the concession of a reduced tariff, and the matter of postal stations was dropped.[57] In 1863, however, the Russian merchants' guild established, chiefly for its own members, a fortnightly mail service between Kiakhta and Peking with connections to other cities in Mongolia, especially Uliassutai and Kobdo. The mail was carried along with the caravan from Kiakhta to Urga, and by soldiers of the Mongolian banner army from thence to Kalgan. After Kalgan the Russians used the services of the Chinese *hsin-chü* for transmission to Peking and Tientsin. In 1870 the Russian government took over this private service and replaced it with its own postal agencies in Urga, Kalgan, and Tientsin, and began to use stamps. It seems that the Russian postal service was not formalized until a later date. G.T. Bishop maintained that Russia opened "post offices" in Kuldja (Ili), Tarbagatai, Urumchi, and Urga in 1899. In 1896, after the completion of the Trans-Siberian Railroad, a sea route for mail was opened between Vladivostock and Chinese ports, and Russian postal agencies were established in Shanghai, Chefoo, and Hankow.[58]

According to the Additional Treaty of Peking, letters were to be sent from Peking and Kiakhta once a month from each end via the *t'ai-chan*, and parcels every other month from Kiahkta to Peking and once every three months from Peking to Kiakhta. No restriction was imposed, however, on diplomatic correspondence—exchange of mail between the Russian Ministry of Foreign Affairs or the governor of East Siberia on the one hand and the Tsungli Yamen or the Li-fan yüan (Court of Dependencies) on the other. Moreover, for very important matters, both the Russian government and the Russian envoy in Peking were allowed to send specially designated Russina couriers through Chinese territories, but the Chinese government was to be informed twenty-four hours in advance of the courier's pending departure from Peking or his intended entry at the Chinese frontier.[59]

Not long after the conclusion of the agreement for the overland service, friction arose. The imperial agent at Urga, who had jurisdiction over the post route with its twenty-five stations between Kiakhta and Peking, requested in a memorial that the Russian minister in Peking be notified of the violations to the terms of the treaty. It was reported that notwithstanding the stipulation for a monthly service, in the first five months of 1861, more than ten mails were dispatched from Russia to Peking.[60] Moreover, Russian officials and messengers who traveled the post route were said to have little consideration for the men and animals placed at their disposal by the Chinese post stations.

> Recently the Russians have traveled back and forth continuously,, availing themselves of the facilities of the post stations. . . whether it is an official or a soldier who travels, all of them ride in carts and carry baggage. Every time they pass through, one or two hundred animals and more than ten carts are used; generally traveling day and night they cover five or six stations a day [against the normal rate of three stages a day] . . . Furthermore, whenever the Russian couriers pass through, they all claim to have urgent business so as to have a pretext for riding posthaste, which obviously strains the I-chan services.[61]

The Chinese authorities insisted upon strict observance of all the treaty specifications, such as those governing the number and weight of parcels, the route of the couriers, and the proper procedure for dispatch. They were disturbed when they learned that a member of the Russian Ecclesiastical Mission had sent a parcel privately without notifying the proper Chinese officials.[62]

Moreover, despite entreaties to the contrary, Russian couriers were not allowed to travel in China without escort, lest they linger on the way or spy on the countryside.[63] Much of the uneasiness on the part of the Chinese, especially that of the local authorities in Mongolia and Manchuria, stemmed from the fear that Russia harbored secret designs on Chinese territory.[64]

It is interesting to compare the vigilance of the Chinese officials toward Russian couriers and their indifference toward foreign postal establishments in the treaty ports, even though the latter were not based on any treaty agreement. For many years the foreign post offices were allowed to carry on their business undisturbed and no action was taken to check their growth.

# Chapter V

# THE ORIGIN AND DEVELOPMENT OF THE CUSTOMS POST

Sir Robert Hart first came to China in 1854 and served in the British consular service before he joined the Chinese Maritime Customs in 1858.[1] He succeeded H.N. Lay as the inspector-general of Customs in 1863, and for half a century he not only headed the vast Customs service, but also advised the Chinese government on a wide range of matters dealing with foreign countries. He won the confidence of the Yamen ministers through his tact and personality, his fluency in the Chinese language, and his thorough knowledge of treaty obligations.[2] Imbued with the Victorian belief in progress, Hart was convinced that China's salvation lay in reform and modernization along the lines of the technically advanced nations of the West. Since the Customs was assigned the job of handling the legation mail, Hart suggested to the Yamen as early as 1861 that China establish a national post office along Western lines.[3] He did not seem to have any definite plan in mind; presumably the British model could be easily adapted for China. By the mid-nineteenth century the benefits of the penny postage and other British postal reforms were fully evident and sucessfully emulated by other countries. At twenty-six Hart was "brimful of what he believed were progressive ideas, and he was so charged with youthful enthusiasm and energy, and self-confidence that he felt himself fit to tackle any problem," as Commissioner Stanley Wright later remarked.[4] In the early 1860s, however, when the Chinese government was still faced with the problems created by the aftermath of the war and the Taiping and Nien rebellions, Hart's proposal for a national post office did not make much impression on Chinese officials.

## The Tsungli Yamen, the Customs, and the Legation Mail

One of the major gains for the West from the Arrow War of 1858-1860 was the right to maintain permanent representatives in Peking who were to enjoy the privileges usually extended to diplomats in the West. Diplomatic correspondence became one of the privileges mentioned in the Tientsin treaties.[5] The Russian treaty of June 13, 1858, established the Peking-Kiahkta Overland Service and regulations for Russian couriers going through Chinese territories; the American treaty of June 18, 1858, gave the American diplomats the privilege of forwarding official documents addressed to the Chinese government in Peking via the Chinese Official Post.[6] The French treaty of June 27, 1858, mentioned the inviolability of letters and effects of French representatives in China, while the British treaty demanded protection of the same. By the extension of the most-favored-nation clause, Article 4 of the British treaty was inserted, in identical or similar terms, in treaties subsequently made between China and other nations.[7] The text of Article 4 stipulates that:

> Her Majesty's Representative . . . shall . . . have full liberty to send and receive his correspondence to and from any point on the sea coast that he may select; and his letters and effects shall be held sacred and inviolable. He may employ for their transmission special couriers who shall meet with the same protection and facilities for travelling as the persons employed in carrying dispatches for the Imperial Government and generally he shall enjoy the same privileges as are accorded to officers of the same rank by the usage and consent of Western nations.[8]

Since the British already had several postal agencies in China it is not clear whether they were contemplating opening additional ones in the newly opened ports, but Shanghai soon witnessed the establishment of a French postal agency in 1862 and an American agency in 1867.[9]

## Legation Mail and the Winter Overland Service

During most of the year it was a relatively simple matter for the foreign legations in Peking to receive or dispatch their mail via Tientsin, the nearest port, for transmission by steamer to other parts. But from about December to March the approaches to Tientsin harbor were icebound and mail had to be transmitted overland via Chinkiang to Shanghai for transmission abroad or to other ports. In the early 1860s the 778-mile route between Peking and Chinkiang was still beleaguered from time to time by the Nien rebels. The Tsungli Yamen, which was opened in 1861 to handle all foreign affairs,[10] undertook to convey diplomatic correspondence for the legations using government couriers; the correspondence of the Chinese Maritime Customs Inspectorate and its international staff[11] was also carried along with the legation mail.[12] In 1863 when the Inspectorate was transferred from Shanghai to Peking and the inspector general was asked to reside in Peking permanently, the task of collecting and making up the mail for dispatch was entrusted by the Yamen to the Customs.[13]

At first the Customs handed the mail bags to the Yamen for transmission by government couriers, but some time later the Customs engaged couriers of its own. [14] This led to the creation of postal departments at the Inspectorate in Peking and in the Customs house at Chinkiang. [15]

The winter months in North China were regarded by foreign residents in China as a dead season, when no letter or newspaper arrived by steamer from the outside world; hence the regular Peking-Chinkiang service was greatly welcomed. Couriers rode on donkeys or mules and usually covered part of the distance—between Yangchow and Chinkiang, and when the wind was favorable also between Yangchow and Ch'ing-chiang-p'u—by boat on the Grand Canal. [16] Except for occasional hitches, such as when mail bags were stolen by bandits, or when men and animals fell into ditches during the rainy season, the service worked very well.[17] In 1870 the Customs commissioner at Chinkiang, Henry C.J. Kopsch, reported with great satisfaction:

> The fourth season of the overland transport of mails of Peking opened on the 25th December and closed on the 25th February.[18] During the period 9 mails were dispatched by this office and 8 received, the journeys being accomplished with very creditable punctuality. The average time occupied by the couriers in travelling to Peking was 10 days, but it has occasionally been performed in 8½ days, or at the rate of 91½ miles a day.[19]

Possibly because the winter service had so inspired him, Kopsch later became one of the most ardent advocates for a modern post office for China.

After 1865, when the Customs assumed full responsibility for legation mail, the service was not confined to the winter months. Quasi-postal departments were established in the Customs at Tientsin and other ports to handle foreign mail during the open season for steamer service, [20] and a regular weekly courier service was inaugurated between Peking and Tientsin. A mail notice of 1867 shows that the mail from Peking to Tientsin was made up every Thursday at noon, and the mail from Tientsin for Peking every Saturday at two in the afternoon. However, any mail arriving in Tientsin from Europe or America during the first part of the week, if handed to the Customs commissioner, was forwarded to Peking by special courier. [21]

In the beginning the service operated largely on a subscription basis. Thirty taels per season entitled a subscriber to send a bag weighing up to three catties (about four pounds) in each outgoing mail from Peking. Extra costs were divided among subscribers at the end of the season on December 15. Rates for forwarding letters of non-subscribers were as follows: letters weighing up to an ounce, four candarins or tael cents (Tls. 0.04); between one and four ounces, twenty cents; between four and eight ounces, fifty cents; each newspaper, two candarins.[22]

This mail service was available primarily for the use of the diplomatic, consular, and Customs personnel and their families—foreigners residing in China in an official or semi-official capacity. Because of the difficulty of transmitting the mail in winter the benefits of the Peking (Tientsin)—Chinkiang overland service were extended to the general foreign community in Tientsin after January 1868 only on a limited basis. Mail bags limited to ten pounds capacity were addressed to an agent of the Tientsin community in Shanghai for distribution and were transported by the Customs couriers. [23]

## The Ch'ing Government and the Self-Strengthening Movement

From 1861 to 1884 the head of the Tsungli Yamen and the Grand Council was the emperor's uncle, I-hsin, Prince Kung (1833-1898). The prince enjoyed great popularity among foreigners and Chinese alike for his role in the peace negotiations in 1860, for his charm and ability, and, above all, for the full confidence placed in him by the regents, the empresses dowager, Tz'u-an and Tz'u-hsi. [24] With the support of a number of brilliant officials Kung was able to restore the prestige of the imperial government, even showing some promise for a glorious restoration and a new era. Many Chinese scholars considered the situation in 1860 as unprecedented in the last 3,000 years of Chinese history.[25] Many Westerners also urged China to change and modernize. [26] Prince Kung and some of his colleagues in the Yamen through frequent contacts with Westerners gradually came to accept the idea that China must strengthen herself by adopting Western technology and methods for the sake of self-preservation. In 1865, two years after Hart became inspector-general, a memorandum entitled "The Spectator's View" was presented to the Yamen with suggestions for introducing railroads, steam navigation, and the telegraph as means of improving conditions for foreign trade and foreign relations.[27] In 1866 the British minister, Sir Rutherford Alcock, also addressed a memorandum to the Yamen; it was drafted by Thomas Wade, then Chinese secretary at the British legation.[28] The Yamen ministers sent out to the governors-general and governors a circular request for comments on the memoranda and their views on the impending treaty revision. Most replies emphasized the urgency of political and military reforms; some advocated the adoption of Western arms and technology.[29] A number of high officials in the provinces—the governors-general Tseng Kuo-fan, Tso Tsung-t'ang, Li Hung-chang, and several others—persistently and energetically pursued an enlightened and progressive policy and, with the Yamen's support and imperial sanction, initiated a variety of projects such as arsenals, shipyards, colleges for interpreters, and sending students abroad.[30] The T'ung-chih reign (1862-1874) witnessed a considerable amount of restoration and modernization, but among the modes of modern communication only the steamship was accepted. Although today a national post office seems to be taken for granted as an indispensable part of modern society, a post office was not even mentioned for official consideration as were the railroads and the telegraph. Although neither memorandum made any special reference to postal reforms both Hart and Wade had constantly urged China to modernize her means of communication. As Hart came to know and understand Chinese politics and Chinese society better, he became increasingly aware of the obstacles which had to be overcome before China could have a national postal system. He was not eager to assume the task until he had clear support from the Chinese government, yet he did not miss an opportunity to renew his proposal.

There are a number of reasons why Prince Kung and his colleagues may have ignored Hart's proposal for a post office. First, by premodern standards the I-chan and the *hsin-chü* seemed to meet adequately the communication needs of the government and the people. The Yamen officials saw no immediate necessity to reform the old systems even though they were told that the new system had been yielding handsome revenues in Western countries and was generally beneficial to the people.

Second, ideologically speaking, the self-strengthening movement in the sixties and early seventies focused primarily on political and military matters; economic enterprises were only a secondary consideration. At that time the post office idea was not widely known and none of the advocates of reform had yet paid any attention to the subject.

Third, the ultra-conservative forces in Peking with backing from high personages blindly opposed any suggestion for change. For instance, when Grand Secretary Wo-jen made a fuss over the appointment of a professor of astronomy and mathematics at the T'ung-wen kuan because the man was a foreigner, public opinion was so influenced that for a time applications for admission to the T'ung-wen kuan sharply decreased.[31] The reactionaries would have succeeded in suspending the shipbuilding project in Fukien had not Tso Tsung-t'ang and Li Hung-chang pleaded vigorously against such a move. Prince Kung and his colleagues were accused of being subservient to foreigners, and Kung's half brother, I-huan, Prince Ch'un, even sent a secret memorial denouncing the foreign policy of Kung and his colleagues and deploring the amount of power in their hands.[32] The Yamen ministers often felt frustrated and had to be very circumspect in their actions, concentrating their efforts on urgent problems. Besides opposition from the reactionaries, there were other obstacles. Sick in heart and body, Wen-hsiang (1818-1876), Pring Kung's ablest and closest colleague, analyzed the situation as he saw it in 1874:

> There is no one who does not talk about self-strengthening, but in over ten years little has been accomplished. The causes lie in the fact that those who despise and disregard foreign affairs rely on empty words, having nothing practical [to offer], [and] those accustomed to the peace are anxious that nothing happens for fear of arousing suspicion. There may be some people who devote themselves to the careful study of current affairs, but owing to the lack of funds, nothing can be achieved or developed.[33]

Lastly, the personal antagonism between Empress Dowager Tz'u-hsi and Prince Kung stifled whatever initiative the latter might have had regarding modernization. In 1865, after the danger from rebellions and foreign wars seemed to have passed, the empress dowager, who had relied so heavily on Kung for her ascent to power in 1860, began to fear Kung's growing popularity and to become irritated at his outspoken manner. She decided to crush him. On the pretext of an impeachment brought by a minor official, Prince Kung was promptly relieved of all his offices and stripped of his honors by an imperial decree which Tz'u-hsi herself drafted. The charges were not substantiated, and the severe punishment caused astonishment and consternation to all. Princes, imperial clansmen, and high officials immediately petitioned the throne on Kung's behalf; public opinion was unanimously and strongly on the prince's side. Since Kung was highly esteemed by the foreign powers and since a change of top personnel in the Yamen might affect China's position, the empress dowager finally revoked her decision and reinstated the prince in all his offices, but the title of "prince counselor," bestowed on him in 1861, was never restored. [34] On the surface the empress dowager maintained a cordial relationship with Prince Kung, but her feelings toward him became increasingly hostile.

In 1874 Prince Kung suffered another humiliation. The behavior of the young emperor, who had just come of age the previous year, began to cause alarm. He spent more time supervising the rebuilding of the Summer Palace than in his studies and state affairs. He was also said to have made visits incognito to the city with eunuchs. Since no remonstrance from his tutors or high officials produced any effect, Prince Kung addressed a memorial to the emperor in which he emphasized the impoverished state of the country, the need for economy, and the emperor's sacred duties. [35] After a stormy audience, during which other princes and officials joined in the exhortation, Prince Kung was degraded and his case handed over to the Imperial Clan Court. The emperor would have liked to reduce his uncle to a commoner if Wen-hsiang had not interceded persistently. When the emperor summarily dismissed Prince Kung, and several other princes and officials, including Prince Ch'un, the empresses dowager intervened in haste. Tearfully they consoled Prince Kung, saying, "For the last ten years if not for Prince Kung, how could [we] have [seen] days like these. His majesty is young and inexperienced; let yesterday's edict be withdrawn." [36]

As a result of this episode, work on the Summer Palace, which was becoming more and more expensive, was suspended; instead, minor repairs were to be made on the imperial gardens. Tz'u-hsi, who was just as enthusiastic about the Summer Palace as was the emperor, probably resented the suspension.[37] As much as she realized the necessity of keeping Prince Kung in office, her coolness toward him gradually turned into hatred unabated by time.[38]

The death of Emperor Mu-tsung in 1874 brought a child cousin to the throne and the return of the empresses dowager to the regency. In 1881, when the senior empress dowager, Tz'u-an, died, Tz'u-hsi became practically the sole ruler of China until her death in 1908. Kung continued to lead the Grand Council and the Yamen, but a subtle change had come over the once energetic and courageous man. He appeared to be vacillating and devoid of initiative, and his health began to fail. When the crisis over Annam developed in 1882-1883, he was absent from his office for almost six months. His failure to formulate a constructive policy led to his dismissal in 1884, but his spirit seemed to have been broken long before. [39]  In the circumstances, the Yamen was very reluctant to take on extra responsibilities unless faced with urgent need for action.

## The Margary Affair and Hart's Proposal in 1876

An opportunity for China to establish a modern post office occurred during the negotiations over the Margary Affair. In February 1875 A.R. Margary, a British consular officer in China, was selected by Sir Thomas Wade, the British minister, to act as interpreter for a British expedition exploring the possibility of a trade route from Burma to Yunnan. Margary traveled with a passport from the Tsungli Yamen, and all went well until he was killed while recrossing the frontier into Yunnan. [40]  The governor-general of Yunnan and Kweichow maintained that it was the act of savages, but Wade insisted that the attack was premeditated and that the local authorities had at least connived at it. He complained bitterly to the Yamen about the unfriendliness of the Chinese officials toward foreigners. The Margary incident was to him merely a symptom of China's "exclusivist policy" which must be abandoned if satisfactory relations with other powers were to be maintained. Wade seized the opportunity afforded by the incident to revise once and for all the diplomatic and commercial regulations governing relations between Britain and China, including the inland tariff, likin, trade regulations, audiences, and relations between Chinese and foreign officials.[41]  Negotiations took place in Peking between Wade and the Yamen ministers and between Wade and Li Hung-chang, the governor-general of Chihli and superintendent of trade for the northern ports, whenever Wade was in Tientsin. The issues became more and more involved as the talks continued from the spring of 1875 into the early summer of 1876.[42]  Wade threatened to withdraw his legation, [43]  and a British squadron was at hand in Chinese waters. [44]

Finally, in June 1876, Wade left for Shanghai, declaring that there was little hope for a settlement. Hart, who had assisted in the negotiations, persuaded the Yamen to send him after Wade. [45]  Before his departure from Peking Hart discussed with the Yamen ministers the concessions that China was willing to make with regard to the Margary Affair and to foreign trade. He suggested to them that Wade might be pleased if the Chinese government declared her intention to establish a mint and a post office to show her sincerity in improving foreign relations, because Wade had been hoping that China would modernize both for her own good as well as for producing conditions favorable to trade. [46]  After consulting with Li Hung-chang, the Yamen approved Hart's suggestion and asked him to work out the details with Li in Tientsin. [47]  The specific proposals regarding the mint and the post office were embodied in the Yamen's memorial concerning the progress of the negotiations. Li recorded his conversation with Hart on July 10, 1876, as follows:

Hart: "Would the establishment of a post office be feasible?"

Li: "[Since] there are already many Chinese letter agencies in the different ports and the postage is cheap, there may not be much business [for the post office.]."

Hart: "I am planning to establish these offices in the treaty ports to transmit letters for the people. For instance, the cost of sending a letter, charged according to weight, from Peking or Tientsin to Hong Kong will be as little as twenty cash to seventy, eighty, or a hundred cash. But money and merchandise will not be forwarded. The Chinese *hsin-chü* in various places will be left undisturbed."

Li: "This can be left to the management of the inspector-general . . ." [Li agreed with Hart that the post office would be beneficial to China and yield a revenue for the government, but he did not see why Sir Thomas Wade would be interested in such a project.]

Hart: "In case His Excellency Wade does not like some of my suggestions, I am trying to present several possibilities. The post office and the mint would facilitate transactions between China and other countries and benefit foreign trade." [48]

Li agreed to the establishment of the mint only in principle; he wanted more time for careful study and deliberation. He was willing for Hart to go ahead with the plans for a post office as long as the existing postal systems were left undisturbed.

As a result of Hart's mediation Wade agreed to meet with Li Hung-chang at the summer resort of Chefoo. Li was then appointed imperial commissioner with plenary power for the negotiations. Meanwhile, for some unknown reason, Wade had a falling out with Hart, although only less than a year before that he had paid warm tribute to Hart for the latter's invaluable service in the negotiations. [49]  By the time they gathered at Chefoo in August 1876 the two were "merely on bowing terms." [50]  To Li's surprise Wade did not bring up during the discussion, which lasted several days, any of Hart's proposals relating to the mint or to the post office. Consequently, neither project was mentioned in the Chefoo Convention concluded in September 1876. Only after the agreement was drafted did Wade approach Li for a letter confirming China's intention to open a mint and a post office, but Li did not comply with this request because it lay beyond the scope of the agreement. [51]

In his report to the British foreign secretary, Wade remarked: "I will say that I half regret the loss of opportunity. Neither the mint nor a postal service, however, appeared to me to find a fit place in any of the three sections of my Agreement."[52] It was generally believed that Wade was not in favor of more power being placed in the hands of the inspector-general of the Customs.[53] Some years later, Hart referred to the incident saying that the national post office project was "excluded by a conspiracy of silence."[54]

### The Customs Post

After the Margary crisis Hart decided to do what he could within his power to establish a national post office along Western lines. Since the Customs had been handling correspondence for the foreign legations after 1863 and postal departments had been established in several ports for some time, it seemed best to expand the existing postal operation into a regular postal service pending the decision of the Chinese government to establish a national post office. In the spring of 1878 Hart instructed the Customs commissioner in Tientsin, Gustave Detring (1842-1913) [55] to begin the experiment in Tientsin, Peking, and the three ports of Newchuang, Chefoo, and Shanghai with a view to future expansion. On March 23 the mounted courier service between Tientsin and Peking, hitherto running several times a week, began a daily run.[56] Detring informed Li Hung-chang that the Customs postal service would be opened to the public on May 1 of that year.[57]

The governor-general's blessing was important, and through his influence it was arranged that ships of the China Merchants Steam Navigation Company opened in 1872 were to carry the Customs mail free of charge. Li also ordered the commanders of Chinese warships in northern ports to report their departures to the Customs so that mail could be dispatched with them.

From December to February (March for Newchuang) practically all mail was forwarded overland. The existing winter service between Tientsin and

Table 13

Schedule of the Winter Overland Service

| Routes | Transit Time (days) | Frequency |
|--------|---------------------|-----------|
| Peking-Tientsin | 1 | daily |
| Tientsin-Chinkiang | 9 | 3 times/week |
| Tientsin-Newchuang | 8 | weekly |
| Tientsin-Chefoo | 12 | weekly |
| Chinkiang-Chefoo | 12 | weekly |
| Chinkiang-Shanghai | 1* | daily |

Source: *Report of the Chinese Post Office 1921;* Shanghai: Ministry of Communication,
Directorate General of Posts, 1922), p. 105.

*Actually about eighteen hours. See letter of Father Em. Becker to Father Hammon,
Aug. 28, 1899, *Chine et Ceylan,* 1.4:300 (1898-1900).

Chinkiang was extended from twice a week to thrice a week. Branch services
were established, northward to Newchuang via Shanhaikuan, eastward to
Chefoo via Tsinan. In the north Tientsin was linked by daily service to Peking;
in the south Chinkiang was linked by steamer to Shanghai.[58]

Distribution and delivery varied according to local conditions so as to
keep expenses to a minimum. For instance, residents of Peking who had agents
in Tientsin would receive their mail through their agents; the Customs Post
would forward mail only for those who had no agents in Tientsin.[59] At
Chefoo letters had to be called for at the Customs house.[60] On the other
hand, in Shanghai, where the municipal council maintained a Local Post, arrangements were made for the latter to distribute the Customs mail gratuitous-
ly. In return the Customs postal departments acted as agents for the Local
Post in areas where the latter had none of its own.[61]

Chinese-language mail was entrusted to native letter agencies. To differ-
entiate them from the regular agencies they were called "Chinese and foreign
letter agencies" (*Hua-yang shu-hsin kuan*). They collected and distributed
mail, but left transmission to the Customs. They charged their own tariffs,
retained the postage, paid their own staff, and remained financially independent of the Customs.[62]

Table 14

Postage Rates of the Customs Post

| Kind of mail | Destination | Postage (candarins) |
|---|---|---|
| A. Domestic | | |
| letters (½ oz. or under) | Peking | 3 |
| | All treaty ports* | 3 |
| | Pakhoi | 6 |
| | Kiungchow | 6 |
| newspapers (1 oz. or under) | | 3-4 |
| B. Overseas | | |
| letters | Hong Kong | 6 |
| | Korea | 6 |
| | Japan | 6 |
| | U.S.A. (via Japan) | 6 |
| | Europe (most countries) | 9 |
| | Central America | 22 |
| | South America | 22 |

Source: *Report of the Chinese Post Office 1921*, appendix B, p. 106.

*Except Pakhoi and Kiungchow.

The Customs Post issued its own stamps. The first issue in 1878 consisted of three denominations: one candarin (green), three candarin (brownish-red), and five candarin (orange), all bearing the design of an imperial dragon among clouds. [63] Postage was charged by weight and a more or less uniform rate was adopted for domestic mail. Senders of letters to places outside China were encouraged to affix the necessary amount of foreign stamps for transmission abroad. Customs stamps were required only for inland postage.[64]

The four issues of stamps in the decade (1878, 1882, 1883, and 1885) were all of the same denominations and generally of the same design. In 1894, in honor of Empress Tz'u-hsi's sixtieth birthday, nine different stamps were designed, bearing symbols of longevity and happiness—the character *shou*, the five bats, the immortelle, peonies—as well as the imperial dragon and characteristic Chinese designs such as the trigram, the junk on the Yangtze, and the carp, the messenger fish.[65] The set was created with considerable ingenuity and care. Hart had hoped thus to arouse interest in the modern postal service. Unfortunately, the war with Japan overshadowed events, including the birthday celebrations and the special issue of stamps.[66]

Li Hung-chang, who had hitherto shown a mild interest in the post office, became quite enthusiastic. A few months after the experiment began in Tientsin he reported to the Yamen that he had observed so far no ill effects and no interference with the I-chan or *hsin-chü*. Li thought that the Japanese minister's charges that the "high postage rates" and the "inefficiency" of the Customs postal service was motivated by the fear of China having her own post office and barring foreign powers from establishing additional postal agencies in China. Li also forwarded to the Yamen a letter from Ho Ju-chang, the Chinese minister to Japan, supporting the post office project. As much as Li seemed to be fully committed to the idea of a national post office, he did not really understand the nature of such an undertaking. He concluded the report to the Yamen as follows:

In my opinion, for the present it may be desirable that the inspector-general be allowed to manage [the postal work], then in the future when some results are achieved and expansion becomes advisable, the Yamen may decide for the local authorities or persons specially appointed to carry on the work along the lines of the present regulations. Perhaps, on Inspector-General Hart's return, you may want to confer with him and ask the superintendents of trade of the northern and southern ports, for reasons of proximity, to supervise the project. I await your decision.[67]

The Customs postal service proved to be regular and convenient. Indeed it seemed so successful even in 1878 that China was formally invited to join the Universal Postal Union. Hart also discussed with the British and French authorities the possibility of withdrawing their postal agencies in China and the future of local posts in foreign settlements, but no further step was taken at that time as the Chinese government was not ready to assume full responsibility for a national post.[68] Nevertheless, encouraged by the results of the postal experiment, Hart announced on December 22, 1879, that he had decided on its continuation and gradual extension to other ports. He appointed Commissioner Detring at Tientsin to act concurrently as commissioner for postal matters with the responsibility of regulating the work of postal departments in other ports.[69] The Customs Post was formally established and designated in Chinese as the Haikuan Po-ssu-ta shu-hsin-kuan (Post Office of the Maritime Customs). Both foreign and Chinese mail were to be handled alike. Relations with the Chinese native letter agencies, the Hua-yang shu-hsin-kuan, were severed, but some of their staff were absorbed into the Customs postal departments.[70]

From the beginning Hart stressed strict economy, He was particularly gratified that the Customs commissioners succeeded in managing the Post in addition to their regular duties without detriment to the efficiency of the revenue service or increase of current expenditure.[71] The Customs afforded the postal departments all the resources at its disposal. With the exception of couriers and letter-carriers, relatively few extra employees were taken on for postal work. In 1879 forty-two couriers were maintained for the main line of the Tientsin-Chinkiang overland service, and the three-month run cost less than Tls. 1,500.[72] Even some twenty years after the inauguration of the Customs Post, in 1896, the postal staff at the central office in Tientsin consisted of one foreign officer, one Chinese clerk, two handymen, seven letter-carriers and three couriers.[73]

Although the Chinese government had allocated no special funds for the work, after December 1879 when the continuation of the Customs postal operation was decided upon, the postal departments kept separate accounts from the rest of the Customs. Only salaries and wages of those taken on specifically for postal work were entered in the postal accounts. Moreover, not only were all receipts and expenses of the postal departments recorded in a cash book—with vouchers in duplicate for every item of expenditure exceeding Haikuan Tls. 4 (six dollars), but quarterly returns were also made to the central office.[74] Western bookkeeping methods were introduced into the postal service as they were for the Customs, and every attempt was made to eliminate graft and dishonesty.

The amount of business that passed through the Customs Post was limited. The Chinese public either did not know of the Post's existence, or, in spite of it, preferred to continue their patronage of the native *hsin-chü*. Most patrons of the Customs Post were Westerners who appreciated specially the winter overland service. In the large treaty ports such as Shanghai or Hankow

the Customs Post encountered competition from foreign postal agencies
(Tientsin being the exception); in the quieter ports the letter-carrying business was hardly profitable. For instance, around 1890 there were 200 to 370
foreign nationals in and around Chefoo (pop. 32,000). Only 2,000 to 3,000
letters a year passed through the Customs Post, a large portion of which came
from the Shanghai Local Post for distribution. Most residents used the Customs
Post for important letters only, because they could send letters by any steamer
at the port free of charge, even though there might be risk of delay or loss. On
the whole more stamps were sold to summer visitors in Chefoo than to residents. The sale of stamps had risen from about Haikuan Tls. 276 in 1883 to
Tls. 462 in 1891, while the salaries of the postal staff averaged Tls. 800 a
year.[75] As late at 1891 the sale of stamps in Newchuang was only half of
that in Chefoo.[76]

In spite of various difficulties the Customs Post was extended to ports
north of Fukien in 1882 and gradually to other ports whenever it became
feasible to do so. The Customs experiment was a significant step toward the
adoption of a modernized national postal system. New postal principles and
practices such as a uniform postage based on weight, prepayment by adhesive
stamps, and distinction between letters and printed matter paved the way for
a national post office. The chief credit should go to Sir Robert Hart in making
this modest beginning without much extra cost to the government or dislo-
cation of the existing postal services. He was careful to leave vested interests
untouched, lest strong opposition hinder progress of the Customs Post. Hart's
contribution is the more remarkable considering his onerous duties as chief of
the expanding Customs service and advisor to the Yamen in so many transactions and negotiations with foreign countries.

## Chapter VI

## ATTEMPTS TO CREATE A NATIONAL POST OFFICE

Almost twenty years elapsed between the formal establishment of the Customs Post and that of the Imperial Post Office. During this time, while the confused political situation retarded modernization, diplomatic and military reverses aroused considerable intellectual ferment and interest in reform which the government could not ignore. Public attention was finally drawn to the need for a new postal system; in the meantime, several postal innovations, including the Customs Post, pointed the way toward the creation of a new institution.

After the dismissal of Prince Kung in 1884 all the Yamen ministers were eventually replaced. I-k'uang, Prince Ch'ing (1836-1916), became the leading member of the Tsungli Yamen; Prince Ch'un became the head of the Grand Council, and by special decree, was to be consulted on all important affairs. Henceforth the Tsungli Yamen and the Grand Council were no longer headed by the same person.

Prince Ch'un, who was generally regarded as a "bitter foe of foreigners and foreign ways," and members of the new government, most of whom were of the same persuasion, surprised the world by their friendliness toward foreign representatives and their interest in reform. They were hailed by the *North China Herald* as "really progressive men," and foreign legations reportedly found them "a great improvement on the old lot."[1] After peace was concluded with France in 1885, the ministers relaxed once more, and the same paper reported that the policy of the government seemed to be going in the opposite direction of what was earlier expected by the world, and the "golden dream of reform" had vanished.[2]

With few exceptions the Yamen ministers were still fairly conservative, but they were not entirely unaffected by the changes of the time. The self-strengthening program eventually broke down many prejudices; closer contacts with the West, exchange of diplomatic missions, sending students abroad, translation of Western works, together with the growth of newspapers brought new ideas and knowledge to an ever-widening circle of people. The conveniences of steamships, railways, and the telegraph replaced fear of Western techniques with appreciation; the clamor for modernization was gradually heard over the voice of its opponents.

The new attitude among the literati may be illustrated by Prince Ch'un's enthusiasm for the new navy and his support for industrialization projects initiated by Li Hung-chang, Chang Chih-tung, and several other prominent governors-general. China's defeat at the hands of France in 1884-1885 stimulated discussions on reform throughout the country. Even if the Chinese ministers were not ready to take active measures, they were willing at least to consider proposals that claimed to render China "strong and prosperous." It was in this atmosphere that the proposal for a national post office was presented again and again to the Yamen officials.

*The* Wen-pao-chü

As time passed the Ch'ing Official Post could not meet all the needs that arose with the new situation, and temporary measures were taken; the *wen-pao-chü* (bureaus for official dispatches) were a case in point.

The Margary Affair, which led indirectly to the formal inauguration of the Customs Post, also brought about two other developments—permanent Chinese representation abroad and the creation of a bureau for official dispatches. At the end of 1876 China sent a mission to England to express regret for the death of Margary. This became the first permanent diplomatic mission abroad, and in 1877 the envoy to England, Kuo Sung-t'ao, (1818-1891),[3] was appointed concurrently minister to France. Since the I-chan only handled documents within China, Kuo arranged with an officer of the China Merchants Steam Navigation Company to supervise the transmission of his correspondence.[4] Major Huang Hui-ho, who had studied in England for several years, was chosen for the task, and an office known as the *wen-pao-chü* was attached to the shipping company in Shanghai.

Following the establishment of a legation in England, legations were set up in France, Germany, the United States, Peru, Spain, Japan, and other countries.[5] Each minister made his own arrangements for transmitting correspondence, resulting in confusion and waste.[6] Finally, consultation between the Tsungli Yamen, the superintendents of trade for the northern and southern ports, and the envoys abroad led to the organization of the Central Bureau for Official Dispatches (Wen-pao tsung-chü) in 1878. The office was located in

Shanghai and was to supervise all outgoing and incoming mail for all the Chinese diplomatic missions that shared the expenses of the bureau.[7] Major Huang continued to head the office, and his two assistants were chosen from among the officers working for the other envoys.[8] To facilitate the transmission of documents between Peking and Shanghai, a sub-office was established in Tientsin.[9]

The central bureau was attached to the China Merchants Steam Navigation Company, and its organization was based on the regulations that Kuo Sung-t'ao had originally formulated:[10]

(1) All documents must be registered, giving origin, dates of arrival or dispatch, the name of the steamer on which they were sent, and other details. Should any article be lost, the sender was to be notified immediately.

(2) Only the head of the bureau, Major Huang, was allowed to have clerks and letter-carriers.

(3) Exact accounts were to be rendered and receipts were to be obtained for telegraph and postal fees paid.

(4) The bureau was to furnish Chinese newspapers from Shanghai to all the legations abroad and make accurate reports of important events not mentioned in the paper.

(5) Letters and packages to families of members of the different legations would go free if enclosed in diplomatic bags; otherwise, postage would be charged to the individual or family concerned. In special circumstances the bureau was authorized to pay postage due for redirection to or from inland China up to 1,000 cash per cover for private mail.[11]

The *wen-pao-chü* proved to be satisfactory and the personnel efficient. In 1881 Major Huang and his colleagues were recommended for promotion by Li Hung-chang and the superintendent of trade for the southern ports for having given complete satisfaction in their performance.[12] The efficiency of the Shanghai and Tientsin *wen-pao-chü* inspired many high officials to establish similar bureaus for their own documents and telegraphs. Such bureaus appeared in Hankow, Canton, Foochow, Amoy, Swatow, Kiungchow, Tainan, and other cities where steamer service was available.[13] The bureau's main duty was to supervise delivery and dispatch; transmission was entrusted either to the China Merchants Steam Navigation Company for domestic circulation, or to foreign postal agencies or shipping companies for circulation abroad.[14]

The bureau often cooperated with the Customs Post and let the latter dispatch its mail.[15] The advantage of these bureaus over the I-chan was quickly recognized, and *wen-pao-chü* continued to be installed in different parts of the country until almost the end of the dynasty.

In the last part of the nineteenth century, some of the I-chan postal functions were assumed by the *wen-pao-chü* or the telegraph offices. Desuetude or infrequent usage caused further neglect and accelerated the deterioration of the service. Local authorities made no attempt to modify the ancient methods of transmission, and the Board of War gave no thought to improving the service. Postal laws became inoperative, notwithstanding efforts from above to enforce them. For example, many provinces failed year after year to submit their annual financial reports on the I-chan to Peking even though edicts were repeatedly issued commanding governors-general and governors to do so.[16] The Board of War, which audited accounts, had no jurisdiction over the local magistrates in the management of I-chan funds. It could merely memorialize and seek support from the governors-general and governors. In 1894 an imperial decree instructed the Board of War to circulate a notice to all these officials to reform the I-chan service.[17]

But "reform" at that time was understood to mean little more than a stricter observance of the laws and regulations, tighter control and supervision by superior officials over their subordinates, the elimination or reduction of corruption, and possibly the purchase of a few good horses for the stations; it called for no innovation and little change. There was nothing basically different in the travel service provided by the I-chan at the beginning and at the end of the century. Officials traveling on special missions at government expenses followed the same itineraries.

In 1891 Ch'ü Hung-chi left Peking to preside over the triennial examination in Fukien, following the eastern I-chan route through Chihli, Shantung, Kiangsu, and Chekiang. He found that some stations used mechanized boats to tow native crafts on the canal and rivers,[18] but the journey from Yangchow to Foochow, most of it overland, took thirty-four days. At the end of the examination he was appointed education commissioner in Szechwan and was able to travel more freely, without conforming to the prescribed route. He covered the same distance by steamer in less than a week.[19] In 1901 Lü P'ei-fen and Hua Hsüeh-lan, examiners for the Kweichow examination, left Peking on June

30 and arrived at Kweiyang on September 10, having traveled seventy-three days with twelve days of stopovers. They managed to shorten their trip from Peking to Paoting from four days to one by taking the train, but they rode in sedan-chairs through Chihli and Honan and changed to boats in Hupeh and Hunan. If they had gone by steamer via Shanghai and the Yangtze, they could have traveled in greater comfort and reached their destination in half the time.[20]

While the I-chan system had remained much the same, the quality of the service had altered considerably by the last decade of the century. Although Ch'ü Hung-chi had noted in 1891 that some post houses on the most traveled routes were in disrepair and wrangles over cart rentals often delayed his departures, and that he had endured some discomforts, there was by and large no unpleasantness in his relations with the local officials.[21] In 1894, however, when Yen Hsiu, the education commissioner for Kweichow, journeyed from Peking to Kweiyang to take up his appointment, and in 1896, on his way back to Peking, he was appalled at the corruption and petty annoyances he encountered at the postal stations.[22] When Hua Hsüeh-lan and Lü P'ei-fen traveled the same route a few years later, the situation was even worse.

After the Boxer rebellion there seemed to be a perceptible diminution of the prestige of the imperial court and of officials from Peking, and a corresponding lowering of morale on the part of local officials. The I-chan stations reflected sadly the general state of affairs. Hua recorded more fully than most examiners who kept diaries in similar circumstances the various incidents on his journey. The unpleasantness must have made a deep impression on him. Owing to the shortage of horses or porters, he and Lü often had their departures delayed, and their lodgings frequently left much to be desired. At one station in Kweichow, they arrived behind schedule and were served a meal prepared two days before.[23] It is interesting to note that as much as Hua and Lü had conducted themselves correctly, their servants were not above asking "gift money" from local governments. Even though he knew it was an evil practice, Hua did nothing to stop it.[24]

On their return trip to Peking Hua and Lü decided that it would be more satisfactory for them to ask the I-chan stations to commute their services into money, and for them to look after their own transportation. However, this involved negotiations with local magistrates for the sums commuted, and more than once Hua and Lü found themselves in rather embarrassing situations.[25]

Traveling was a very complicated affair. Hua rode in a sedan-chair attended by five servants on horseback. They set out from Peking across the northern plains with eight porters, eight horses, and two or three carts. Carts were not suitable for traveling in the south, so the luggage had to be carried. In Hupeh Hua and Lü and their attendants had 70 men serving as chair-bearers and porters, and in Kweichow 135 men. On their return trip they had as many as 212 men and ten pack-animals in northwest Hunan, and their combined baggage was carried in thirty to sixty loads. [26] To compound their difficulties, they found that when they reached the north that the court was also returning to Peking from its exile, and all available carts, horses, ponies, and men were pressed into service. [27]

If the I-chan service gradually deteriorated in China proper, conditions of post stations in Sinkiang and Mongolia were much worse. It was not very clear whether local governments were always quite scrupulous in observing the postal regulations and in keeping up the facilities at the stations, but there is abundant evidence that many travelers abused their privileges. In 1890 the military governor of Uliassutai was impeached for allowing his retinue to "plunder" the stations on his way to take up his post. [28] K'uei-pin, the lieutenant military governor of the Altai post road, was asked to make a report on the general conditions of the t'ai-chan. It was found that the new military governor and his household had used as many as eighty-eight to a hundred camels—thirty to forty was the usual number for a high official; his subaltern, Chi-t'ung, who came to meet him together with the governor's servants, extorted money, horses, and sheep from the stations by intimidation and even physical violence. Sometimes they wanted the animals commuted into money payment, and netted on this trip some 3,000 to 4,000 taels. They were said to have taken money from merchants and in return to have included them in the governor's household so that they could enjoy the facilities of the official postal stations, but evidence for these charges was difficult to obtain. The military governor of Uliassutai was subsequently removed from office, the money and goods confiscated, the subaltern sentenced to hard labor in Sinkiang, and fourteen others were also convicted. [29]

K'uei-pin also memorialized that the Mongolian postal stations were very much abused by officials and garrisons that traveled back and forth along these routes. Local stations were often pressured for money and animals. The worst abuses were to utilize the services of the stations for all kinds of errands. This became a heavy burden on the people and a drain on the economy of Mongolia. [30]

### Domestic and Foreign Postal Agencies

The commercial letter agencies, unlike the official postal system bound by tradition and government statutes, were much more flexible and quick to seize the many opportunities offered by improved transportation facilities to transmit their mail by steamer and railroad whenever possible. During the 1880s railroads gradually extended from Tientsin and Peking eastward to Shanhai-kuan and Manchuria and southward to Nanking and Shanghai. Business for the *(min) hsin-chü* flourished considerably with improved transmission, and many of them were able to reduce their rates, adopting almost a low uniform postage (see rates for Kiukiang agencies in 1891 and 1901 in tables 9 and 10). As a group the letter agencies were becoming a force to be reckoned with.

The foreign postal establishments too became increasingly active. In 1885 Germany joined the other nations in extending her mail line to China. The letter-carrying business in China was indeed a "free-for-all." Local posts of foreign settlements mushroomed and issued stamps almost indiscriminately. It was the extraordinary activities of the foreign postal establishments that prompted Hart and many other Customs officials to push forward the project for a Chinese national post office before the situation became too chaotic.

During the last quarter of the nineteenth century China had seven or eight types of postal service, namely, the Ch'ing Official Post consisting of the I-chan, the t'ai-chan, and the P'u, the *hsin-chü*, the Customs Post, the central and provincial bureaus for official dispatches, the postal agencies es-tablished by foreign governments in the treaty ports, and the local posts es-tablished by municipal councils in foreign settlements—each independent of the other. Although postal services had multiplied, many places in the interior, such as Yunnan and Kweichow, were still in need of adequate postal facilities. Another decade was yet to pass before the attempts to create a national post office were successful.

### The Taiwan Post Office, 1888-1895

The post office established in Taiwan by its energetic governor Liu Ming-ch'uan (1836-1896) was an interesting example of indigenous reform efforts.

The island became a separate province in 1887, and Liu's program of modernization included a railway from Taipei to Keelung, telegraph lines from north to south, cables to Fukien and the Pescadores, and a steamship line between Taiwan, the China mainland, India, and the South Seas. [31] Under Liu's auspices the Taiwan Post Office was established on March 22, 1888. Some fifty stations of the existing I-chan system were reorganized, each under a postal clerk, and the whole system under a circuit intendant. [32]

The post office operated along Western lines; postage was based on weight and distance and prepayment made by means of stamps. There were two kinds of stamps: Taiwan stamps (Taiwan *yu-p'iao*) for franking official dispatches and commercial postal stamps (*yu-cheng shang-p'iao*) for public use. At first the stamps were printed locally on thin Chinese paper with blank spaces for the date, the weight, and the postage, but they were not adhesive. Each stamp had a tab with a number and was not sold previous to posting. This made every letter practically a registered article. Mail was delivered by foot-carriers twice a day, except to mountain passes or places difficult to reach. [33]

During its first year the Taiwan Post Office compared favorably with the I-chan in efficiency and economy. There was no delay or loss of official documents, according to the governor's report, and expenses for the year 1888-1889 came to about Tls. 10,000 against Tls. 15,000 to Tls. 16,000 for the years before the reform. [34] The experiment was shortlived, however, for in 1895 Taiwan was ceded to Japan.

*Li Kuei's Proposal, 1885*

In 1885 the proposal for the establishment of a national post office was formally presented to the Tsungli Yamen through the regular channel of Chinese officialdom. [35] Three individuals who were in favor of introducing a Western postal system to China happened to be living in Ningpo at the same time, namely, Hsüeh Fu-ch'eng (1838-1894), intendant of the Ningpo and Shaohing circuit, [36] Commissioner Henry Kopsch, an Englishman who had previously supervised the Customs winter overland service between Chinkiang and Tientsin, [37] and Li Kuei, a writer in the Ningpo Customs who had joined the service in 1865 and had accompanied the Chinese mission to the Centennial Exhibition at Philadelphia in 1876. In 1885 Li presented Hsüeh Fu-ch'eng with a proposal for the establishment of a national post office and furnished

him with a translation of postal regulations at Hong Kong for reference.[38]
His proposal is no longer available, but it is said to have mentioned that letters from Chinese merchants and laborers overseas were sometimes held up
for ten years for lack of postal arrangements between China and foreign
countries.[39]

Hsüeh, who had hinted in one of his essays at the eventual replacement
of the I-chan by the modern postal system when railways spread in China, took
up Li's proposal seriously and consulted Henry Kopsch on the subject.[40]
Kopsch subsequently drew up a memorandum which, together with Li's proposal, was presented by Hsüeh to Tseng Kuo-ch'üan (1824-1890), the governor-general of Liang-Kiang and concurrently superintendent of trade for
the southern ports.[41]

In his memorandum Kopsch set forth the reasons for wanting a Chinese
post office, outlining the plan for its administration and citing the beneficial
effects of the British Post Office to support his case. Owing to the lack of
postal facilities, he said, other nations had established post offices in Chinese
treaty ports. If this state of affairs were allowed to continue China's national
prestige would be affected, for foreign countries considered the speedy transmission of mail for the people a duty of the government. Therefore, to bring
about the withdrawal of foreign post offices, China should without further
delay establish her own post office, and to win the confidence of other countries, the postal service must be very efficiently handled by those knowledge-
able in the subject. He suggested that at the start the new institution might be
attached to the Maritime Customs in the treaty ports, where Westerners were
already on the staff and where the Chinese knew foreign languages. As for
foreign mail, arrangements would have to be made with Hong Kong and Japan
for forwarding it both ways. Kopsch also described briefly the success of the
British postal system and its money order and postal savings services which
were particularly helpful to workers and the poorer people. He claimed that
Britain's postal revenue in 1884 was the equivalent of Tls. 31,924,900, but
the expenses were only Tls. 18,854,000.[42] This memorandum inspired many
others in the decade following 1885.

Tseng received favorably Li's proposal and Kopsch's memorandum and
forwarded them to the Tsungli Yamen for consideration. Without committing
itself the Yamen turned the matter over to Sir Robert Hart and the Shanghai

Customs superintendent for further study; their findings were to be referred back to the superintendents of trade for the northern and southern ports, Li Hung-chang and Tseng Kuo-ch'uan respectively, for consideration. [43] Meanwhile, Kopsch had traveled to Hong Kong to sound out the British authorities on the withdrawal of their packet agency in Shanghai. In the spring of 1890, almost five years after Li Kuei submitted his proposal, Hart was informed by the Tsungli Yamen that it was satisfied with his plan insofar as it seemed not to interfere with the livelihood of the people or to disturb the *hsin-chü*. As a preliminary step toward the establishment of the post office the Yamen authorized the Customs Post to expand to all treaty ports, but no imperial sanction could be obtained until some progress had been made. [44]

What happened during those five years? Apparently, whenever the project was brought up for discussion the Chinese officials raised the question of the livelihood of the Chinese people. [45] It was generally known that mass unemployment resulted from the decision of the Ming government to reduce the number of postmen (*i-tsu*) in the official courier service for reasons of economy in 1629. In Shensi drought and famine had so aggravated the plight of the unemployed that many of them joined the rebels or roving bandits. [46] Historians generally agree that the ranks of the bandits were swelled by destitute postmen and others who had depended on occasional employment in the I-chan for a living—Li Tzu-ch'eng, one of the rebel leaders, was a former postman [47]—but whether this actually hastened the downfall of the Ming has been differently interpreted. [46] Since rebellions and banditry continued into the latter half of the nineteenth centry, it was quite likely that the Ch'ing government was wary of repeating the mistake of the Ming government. [48] If a national post office on Western lines was established, over seventy thousand I-chan employees and almost an equal number of *hsin-chü* employees would be affected.

In 1886 Kopsch was transferred to Shanghai. He had envisaged the development of the Customs Post into a national post office with a special office at Shanghai. In 1891 the report of his successor at Ningpo, [49] Commissioner H.F. Merrill, revealed some of the serious difficulties in such an undertaking:

With much patient labour and study during the years 1884-6, Mr. Kopsch drew up in detail plans embracing the establishment and administration of a national post office with all the functions of the fully developed post office of the day, the absorption of the work of the various Foreign post offices now existing in China, the recognition abroad of China's postage stamps, and the ultimate admission of China into the Postal Union if deemed advisable. These plans, which were approved by high Chinese authorities, encountered some opposition amongst the Foreign mercantile community in Shanghai, and this and the pressure of more urgent business caused them to be laid aside for the time being.[50]

The foreign mercantile community, accustomed to its privileged position in Shanghai—with its own municipal council and police force, independent of Chinese government jurisdiction—would naturally be reluctant to see any of its privileges curtailed. Opposition to the suggestion that its own local post be suppressed is conceivable, especially at a time when the post office business was thriving.[51] Moreover, doubts not necessarily well-founded as to the efficiency and reliability of a Chinese institution and its adequacy to meet foreign needs were often raised as a part of the objection.

Meanwhile, urgent business such as the question of likin, the suppression of smuggling through Hong Kong and Macao, the negotiations with Britain over the Burmese and Tibetan borders, and the increasingly strained relations with Japan over Korea preempted the Yamen's attention. Since 1885 Li Hung-chang, one of the principal supporters of the postal scheme, had also devoted much of his time and energy to the building of a modern navy.[52]

As much as Hart wanted to see China modernized, he had learned to be patient. By this time, as head of an enormous administration and the Yamen's most experienced and trusted advisor in foreign affairs, he was involved not only in diplomatic negotiations concerning trade and the Customs, but also in many other matters such as securing a British instructor for the northern squadron of the new navy.[53] When Kopsch reminded him of the postal project, he replied that he did not want China to enter the Postal Union until the work could be properly carried on, and he was in no hurry to make any advance until he was relatively free to devote adequate attention to the matter.[54] Thus, for a variety of reasons the postal system remained as it was before, except that the Customs Post was to be expanded to all the treaty ports, which would have happened with or without formal authorization from the Chinese officials.

*Hart's Plan, 1892*

Pleas for the establishment of a post office were frequently renewed. The Ningpo and the Shanghai Customs superintendents continued to submit reports to the Tsungli Yamen on the activities of the foreign postal agencies, which, according to the superintendents, thrived on the pretext that the Customs Post had no national status. Toward the end of 1892 Hart urged the Yamen ministers once again to hasten the establishment of an imperial post office; in view of the difficulties already experienced by the Customs Post, he warned them that further delay in instituting an official post would lead to other complications.[55] The immediate cause which prompted Hart's action was the widely discussed project of the Shanghai Municipal Council. Since the Local Post in Shanghai was so prosperous and so efficient, it was suggested that the municipal council should negotiate for its admission as a member of the Universal Postal Union. "If a step of this sort was actually being taken it would be a further encroachment on China's sovereign rights, and would have a pernicious effect on the development cf the Chinese Post Office . . ."[56] Hart therefore instructed his agent in London, James Duncan Campbell, to proceed to Berne and to inform the Postal Union at its headquarters that "Shanghai was not a free city but a Chinese port and had no legal status entitling it to be recognized as having power to negotiate or enter the Postal Union."[57] It was discovered, however, that the municipal council's intended action was not known in Berne. The Swiss Foreign Office gave its assurance th in the event an application should come to the Swiss Federal Council from the Shanghai authorities, no decision would be made without referring to the Chinese government through the inspector-general of Customs.[58]

To encourage the Tsungli Yamen to lay the proposal for a national post office before the throne, Hart presented a plan for the administration of the institution. According to the plan, post offices would first be established in cities along the coast and the Yangtze, then gradually extended inland. It was estimated that in six or seven years no town or large village in the empire would be without a branch of the Imperial Post Office. He also suggested that a European postal superintendent be appointed at each provincial capital and a European postmaster at each prefectural city, "assisted, at first, by English-educated Chinese clerks," while trained Chinese be put in charge of district and smaller post offices.[59]

The Yamen officials were more or less convinced by then of the necessity of a national post office and wanted to see it established, but they could not adopt the plan Hart presented.[60] At the end of March 1893 Hart wrote to Kopsch, who had been eagerly waiting for news of the Yamen's decision, that "The Yamen has retreated into its shell *re* postal extension and its 'go-ahead!' is again postponed."[61] Neither Hart nor the Yamen gave an explanation, but two obstacles apparently stood in the way, the question of personnel and finance

If Hart's plan were carried out, over 300 Westerners[62] would have to be engaged as postmasters by the national post office because there were in the 18 provinces 185 prefectures (*fu*), 72 departments (*chou*), 45 subprefectures (*t'ing*), and 1,303 districts (*hsien*).[63] Many Chinese feared that such a large number of Westerners would increase foreign influence in Chinese administrative machinery.[64] The higher officers of the Maritime Customs were all Westerners, but the Customs had to deal with foreign merchants and were confined to the treaty ports. As much as it was in the Chinese tradition to employ foreigners in government service—"to borrow talents from other countries," to let them control an institution with such vast ramifications as the post office was evidently a different matter. Furthermore, the employment of a large number of Westerners who had to be paid much higher salaries than the natives made the postal project a very expensive undertaking.

The lack of experts in the matter of postal reform was often lamented by the high officials.[65] Aside from Li Kuei there were apparently not many Chinese who were conversant with the intricacies of the Western postal system, and the Yamen relied entirely on Hart and his subordinates to devise a plan and to carry it out. Although Hart was eminently qualified to take on such responsibility he had enough work dealing with the Customs, and his services were frequently required for various diplomatic negotiations which might or might not have to do with trade. To keep himself informed of events at every port he required that every Customs commissioner write him a fortnightly semi-official letter in addition to official dispatches. Hart himself wrote back almost as often, giving advice, encouragement, or warning—whatever was necessary even to the last detail. He knew what went on in each area and was concerned with the families and welfare of his staff. He even declined the appointment to be British minister in Peking in order to continue his service with the Chinese Customs. He was obliged to postpone his home leave more than once. [66] Overworked and in need of a vacation, he was in his den from "dawn until dusk" and seldom went out of his house. [67] The only relaxation he knew was probably his weekly at-homes on the lawn of his spacious garden with facilities for tennis and dancing to a Chinese band playing the latest European music. [68]

From 1875 to 1895 the financial situation of the Ch'ing government showed increasing signs of strain. Although the total national income was much higher compared with that of the first half of the century, expenditures had also grown in greater proportion and at a faster rate. [69] The fall in the exchange value of silver further aggravated Chinese governmental finances, for large sums were being spent on imports of arms, ammunitions, ships, and machinery. [70] In the last decade of the nineteenth century foreign loans were frequently obtained for carrying out industrialization projects. At the same time the extravagance of the imperial court went on unchecked after the pliable Prince Ch'un came to power. [71] While the needs of the government multiplied, the taxable resources were limited; the chief sources of revenue were the land tax, the salt tax, the maritime customs duties, and likin, but an increase of the last two items was restricted by treaties. [72] The situation grew worse as the century advanced. Owing to the depleted reserves in the treasury every emergency, such as the Sino-French War of 1884 and floods in the Yellow River, made it necessary to raise extra funds through new devices, but the results generally fell far short of expectations. [73] The imperial government depended to a certain extent on the provinces for extra funds as well as for the remittance of land taxes, but the provincial governments not only often let regular taxes fall in arrears, but also were not particularly cooperative about raising new revenues. [74]

In the circumstances it is doubtful whether money could be secured for the post office, which was by no means considered a priority project. The I-chan service was being used less and less in several provinces, but the funds allowed for local expenditure still figured in the books because the officials could easily find some other use for the money. Any attempt to reallocate these funds for a new postal service not under the direct control of the local authorities would most likely encounter strong resistance.

When Hart made his suggestion about the post office to the Yamen in the winter of 1892-1893, he was fully aware of the many problems involved. Furthermore, he realized that the more difficult problems were probably more apparent to himself than to others, since he would be the person to carry out the plans. However, he was willing to give it a try. The Yamen ministers were much more cautious. The government was spending vast sums on armaments and on the navy, and the interest in the postal project which Hart and others and tried to kindle was soon extinguished by fiscal and other practical difficulties. Before the postal project was shelved once more Hart wrote to Kopsch saying, "I am in no hurry, being more than ever alive to the expediency of only going where I can go safely."[75] He admitted that the time was not yet ripe. On looking back to the occasion, Commissioner Stanley Wright remarked, "Had this Imperial Edict been issued in 1893, had it . . . made no financial provision for the establishing and upkeep of the national Post Office, Hart would have found himself confronted with the old problem of how to make bricks without any straw."[76]

### Renewal of Pleas, 1893

In January 1893 the Shanghai Local Post boldly ordered the use of Local Post stamps compulsory for local mail as well as for mail in transit to other treaty ports. A high surcharge was often imposed on these stamps. In indignation, European residents in half a dozen ports formed committees to create post offices of their own in the settlements. The Hankow Local Post, which originated in 1877, was the first to issue its own stamps in May 1893; other local posts followed suit.[77]

Kopsch reported these activities to Hart, who had since learned that the Shanghai Local Post would not be admitted as a member of the Universal Postal Union. Since the Yamen was still reluctant to assume the responsibility for creating a post office, Hart wrote to Kopsch on June 28, 1893, saying, "The important thing is to get letters safely and quickly carried, and if China does not care to vindicate her dignity, it does not matter who carries."[78]

It appeared that Kopsch had meanwhile succeeded in persuading the superintendent of the Shanghai Customs, Nieh Ch'i-kuei, to make a move. Nieh sent a petition to the governor-general of Liang-Kiang, Liu K'un-i (1830-1902) to take up the postal project. Sometime between June and July 1893 the Tsungli Yamen received communications from Li Hung-chang and Liu with Nieh's report of the Shanghai Municipal Post and its plan to establish branches at different ports, and they both urged that prompt action be taken about the much discussed national post office. In August 1893 Kopsch also sent Nieh a translation of the regulations of the Universal Postal Union to be forwarded to the Yamen for reference. [79]

Since the postal project had been brought to their attention many times and endorsed by two such senior statesmen as Li and Liu, the Yamen ministers could not very well put it aside without a good reason, and Hart was again asked to "consider, report, and propose." [80] Hart was elated. He wired Kopsch immediately to find out what the rates were for the conveyance of mail and the contribution to the Universal Postal Union [81] and asked his agent in London to proceed once more to Berne to ascertain what steps China had to take in order to enter the Union. [82] By December 1893 Campbell had gathered the necessary information. The procedure for joining the Union was quite simple: The Tsungli Yamen should address a *demande d'ahésion* in an official note to the Swiss Federal Council and inform the Union of the date on which the Chinese Post Office was to be established and the Union regulations were to come into effect. It was pointed out, however, that "entrance into the Postal Union was politically a neutral action and it would not in any sense imply a suppression for the foreign post offices then functioning in China." [83]

No record of Hart's plan is available. Evidently the wiser for his past experiences, he said that he kept many of the postal problems to himself lest they should revive the Yamen's reluctance to proceed and thus postpone action again, perhaps indefinitely. [84] To increase the chances of having his plan accepted, he aimed at a "simple beginning and modest development." He was not in favor of the suggestion that experts from the British Post Office be engaged to help with the undertaking on the ground that newcomers lacked the necessary knowledge of the needs and conditions in China for which allowance had to be made. [85]

The Yamen seemed to have approved his proposal in general, but before presenting it to the throne, had it circulated among the governors-general and governors for their opinions and comments—a current practice then for important matters. [86] In the spring of 1894, before the comments had all come in, the Korean crisis developed and war with Japan seemed imminent. [87] The situation steadily worsened until hostility broke out on August 1. In the circumstances the post office project was the last thing that the Yamen could consider. [88] Here ended the fifth formal attempt—the last of three in a decade—to establish a modern post office in China.

## Chapter VII

## THE ESTABLISHMENT OF THE IMPERIAL POST OFFICE

*Agitation for Postal Reform*

Since 1860 a small but increasing number of Chinese officials and scholars were seriously concerned with the problem of modernization and the possibility of adopting Western technology for China's needs. Apart from Tseng Kuo-fan, Li Hung-chang, Tso Tsung-t'ang, and a number of high officials who were well known for their views and their modernization projects there was Feng Kuei-fen (1800-1874), a scholar who was associated with Tseng and Li in the suppression of the Taiping rebels. He was one of the first to write about the desirability of acquiring Western learning. He wrote so much on the subject that an essay entitled "Ts'ai I-chan i" (On the abolition of the I-chan) was mistakenly attributed to his pen.[1] Actually it was not until after the installation of the telegraph in 1881 that any public notice was taken of the Western postal system, and not until after the mid-1880s that essays on postal reform began to appear.[2] Cheng Kuan-ying,[3] a scholarly merchant who had traveled abroad, and the famous Baptist missionary Timothy Richard[4] described the workings of the Western postal service, and upheld the British post office as a particularly successful example of modernization.

The arguments in favor of adopting a Western postal system for China were as follows: China would have a new source of revenue; she would not have to spend large sums annually on the I-chan and the Official Post. She would also have a legitimate reason for asking the foreign powers to withdraw their post offices from China, and thus vindicate her national dignity. It was pointed out by some essayists that in the West the state had come to assume the duty of providing postal facilities for the people, and Cheng Kuan-ying, who had seen Western post offices in operation, emphasized the special role of postal communication in enlightening the people, one of the basic needs for China to become "prosperous and strong."[5] He advocated strongly that a post office organized along Western lines be established even before the I-chan was abolished. He said that since the development of steamships, railroads, and the telegraph the I-chan had lost much of its importance, and in time it would be superseded entirely by the new methods of communication.[6]

Among Chinese writers who wrote on postal reforms Cheng seemed to be one of the very few who really understood the nature and the organization of the Western postal system. The various plans suggested by the others reflected the influence of the then current formula of "government supervision and merchant management."[7] Some authors suggested that the I-chan adopt Western postal practices; others proposed the establishment of a semi-official institution to be attached to the existing telegraph or steamship offices. In the years immediately preceding the establishment of the Imperial Post Office in 1896, discussion on postal reform frequently appeared in Chinese newspapers which by then had begun to exercise an appreciable influence on the literati. [8]

If the Sino-Japanese War had interrupted the planning for the national post office in 1894, it also had the salutary effect of hastening its establishment. The shock of the defeat was in many ways more humiliating and definitely more widely felt than the British and French occupation of Peking in 1860 or other wars in the century; the self-strengthening movement over a generation proved to be a failure in comparison with the Meiji reforms in Japan, China's erstwhile pupil. The advocates of modernization and Westernization were vindicated, and change became the order of the day.[9] On the crest of the wave of agitation for reforms, the postal reform project was finally completed and placed before the emperor with a request for sanction.

In August 1895, four months after the conclusion of the war, Hart drew the attention of the Tsungli Yamen once more to the plan for the post office.[10] At that time Prince Kung was reinstated as the leading minister in the Grand Council and the Yamen, and Li Hung-tsao (1820-1897) and Weng T'ung-ho (1830-1904), who had been appointed to the Grand Council with Prince Kung, also became members of the Tsungli Yamen. In his position as one of the emperor's tutors, Weng was very influential. He had been interested in progressive ideas since the late 1880s, and China's defeat convinced him that changes in the old order were urgently needed. Prompted by patriotic feeings as well as by personal ambition, he had tried to interest Emperor Te-tsung (Kuang-hsü, 1875-1908) in reforms and to identify himself, at least for a while, with the cause of modernization. [11] Li Hung-tsao was generally considered a conservative, but he had always allied himself with Prince Kung and seemed to have followed the latter's policy on the whole. Hart recorded one of his interviews at the Yamen on the subject of postal reform as follows:

I had a long talk with the new Ministers last Friday—Weng T'ung-ho
and Li Hung-tsao, and they gave me quite a lecture on the unsuitability
of proposals that are not feasible. "Anyone can—everyone does—make
proposals," they say, "but *we* have to work them; *their* part is easy
enough—ours is quite another thing!" [12]

It took the Yamen several months to work out the problems with Hart.
Moreover, delay in replies from the high provincial authorities and the uncertainty of the political situation had also prolonged the process of deliberation. [13] It is not clear who else in the Yamen besides Weng and Li took an active part in the preparation of the postal project, but at least two eminent
elder statesmen were now seriously and conscientiously taking the project in
hand, with a genuine desire to carry it through. With their combined effort,
experience, and caution, and Hart's knowledge of the modern postal system in
the West and his intention of adapting Western methods to Chinese conditions,
they were able to produce eventually a plan practical enough to ensure its
adoption.

## Chang Chih-tung's Memorial

In Liu K'un-i's absence[14] Kopsch had been seeing Chang Chih-tung who
was then acting governor-general of Liang-Kiang. Hart believed that if Chang
were able to take up the postal project he would probably be able to put it
through because he enjoyed great prestige at the time, but Hart preferred to
see the imperial government assume the responsibility. [15] When Kopsch
conferred with Chang in Nanking in September 1895 he acquainted him with
the various preparations and plans already made for the national post office
and pressed him to hasten its establishment. [16] In his former position as governor-general of Liang-Kuang (Kwangtung and Kwangsi) and then of Hu-
Kuang (Hunan and Hupeh), Chang had initiated new schools, mines, mints,
textile mills, ironworks, railways, arsenals, and a host of other modernization
projects. He must have been rather surprised that his attention had not been
drawn to the postal reform project before this. On December 27, 1895, he
presented a memorial proposing the initiation of three practical measures, one
of which was the institution of a national post office modeled after the Western system. [17]

In the memorial Chang summarized all the efforts that had been made
for the establishment of a post office since Li Kuei presented his proposal in

1885, and he repeated all the arguments that had been put forward in support of such an institution. In conclusion he requested that orders be given to instruct Hart to complete the plans and carry them out. On January 3, 1896, Chang received an imperial rescript stating that the postal project was already under consideration by the Yamen and that progress was underway.[18]

In 1895 the judicial commissioner of Kwangsi, Hu Yü-fen, had also memorialized on the desirability of a post office and the abolition of the I-chan. He claimed that the government spent three million taels every year on the I-chan in addition to the large sums spent by the high provincial officials for postal transmission when the I-chan no longer served its original purpose. Except in a few provinces, officials traveled by steamer whenever possible and asked that the traveling expenses be commuted into money payment. While praising the conveniences of the Western postal system, Hu summarized the current views and plans and suggested that special officers be attached to the offices of the China Merchants Steam Navigation Company, the telegraph, or the railroad, to take charge of both official and private mail, first in the treaty ports and railroad stops, then, in conjunction with the *hsin-chü*, throughout the rest of the country.[19] Hu's plan may have been pragmatic, but it was certainly not in keeping with the practices of postal transmission in the West.

## The Memorial of March 20, 1896

Finally, on March 20, 1896 (Kuang-hsü 22.2.7), the Tsungli Yamen presented the famous memorial requesting that imperial sanction be given for the establishment of a national post office along Western lines.[20] The memorial pointed out the merits of a Western postal system and the advantages which China would derive from a modern post office. It explained that in the West the postal service was headed by a high ranking minister and benefited both the state and the people. On the subject of revenue, the memorial said that in 1892 "the United States Post Office alone received an annual income amounting to $64,209,490, and Chang Chih-tung's estimate of British postal revenue at 30 to 40 million taels was only approximate . . . Western postal administrations complemented the telegraph service and utilized trains and steamers for transportation. Recently France had organized a company, and all ten steamers were named mail boats . . ." The report went on to say that some sixty nations had joined the Postal Union, and their membership entitled them to exchange mail speedily, regularly, and safely at very cheap rates over vast distances. There were several million Chinese living in San Francisco, Hawaii, Singapore, Penang, Cuba, Peru, and other places overseas, and they would no longer have difficulty communicating with their families if China became a member of the Union. In the future, the Yamen ministers pointed out, China might have her own vessels carrying mail and native goods abroad.

The memorial also mentioned that the post office project was discussed with the inspector-general of Customs in 1876 during the Margary Affair, long before 1885, as Chang had reported in his memorial. It then traced the origins of the Customs Post and all the plans and suggestions for moderniza-tion made during the two previous decades, the increasing number of foreign post offices in China, and the difficulty of expanding the Customs Post with-out giving it national status.

The completed plan in the form of forty-four articles was prepared by Hart in consultation with the Yamen over a period of six months. The ministers presented it with the following recommendations: (1) After long and extensive inquiries, they had come to the conclusion that the post office project should be seriously considered and immediately under-taken. (2) The plan presented was the result of many years of careful study, experience, and deliberation. It would not interfere with the livelihood of the people. Since details of the plan were still subject to revision and additions when required by circumstances, any ill effects could be quickly eliminated. (3) For the above reasons a national post office should be established by ex-tending the Customs Post from the treaty ports inland, eventually covering the entire country. (4) Management of the post office should be entrusted to Hart, while the Yamen would continue to be generally responsible.

An imperial decree granting the request in the memorial was issued the same day the memorial was presented. The Imperial Post Office (Yu-cheng chü) came into being on March 20, 1896.

## The Regulations of the Imperial Post Office

The forty-four articles in Hart's plan come under four headings: postal transmission between treaty ports, between treaty ports and inland areas, be-tween treaty ports and foreign countries, and general postal regulations. The terms were purposely kept general and simple in order to be easily intelligible to the public as well as to the postal staff. The post office would call for few changes in the existing postal arrangements except a gradual extension of the Customs Post. The management of the post office would remain much the same as that of the Customs Post except that over-all direction which had been in the hands of the commissioner in Tientsin since 1879 would be transferred to the statistical secretary in Shanghai who held concurrently the title of Postal Secretary. [21]

It was hoped that the authority of the Imperial Post Office would be established, even though the principle of government monopoly could only be partially asserted at the time. No mention was made of the Ch'ing Official Post, the I-chan, the P'u, or the *wen-pao-chü*. There was to be little interference with those people whose livelihood depended on the letter-carrying business, but some restrictions were imposed on the *hsin-chü*, bringing them in line with the national institution. They were to register with the Imperial Post Office where there was a branch established or to be established, and in these localities no commercial firm or individual except the registered agencies could engage in postal business. Furthermore, the agencies were not allowed to send their mail by steamer. Only the Imperial Post Office and the registered letter agencies could enjoy the privilege. Apart from these restrictions, the letter agencies would continue as usual, with the Post Office handing over to them some of its mail for distribution in the interior. The letter agencies were allowed to charge their own fees and their business increased considerably as a consequence of the arrangement. As for postal communication with other countries, the Imperial Post Office would continue to send mail abroad through the foreign agencies in China, and in all the transactions observe rates and regulations of the Universal Postal Union.

The Chinese public knew almost nothing or very little about the Western postal system and many viewed the proposed Imperial Post Office with some misgiving. The following comments by a group of officials of the Board of Revenue probably represent a sizable section of public opinion on the innovation:

A: It is simply unheard of that the government should organize a post office to transmit letters for the people . . .

B: I suppose there is no reason why [the government] should not do so.

C: At present there are mounted couriers in every province; why does a post office have to be established? [And] there are letter agencies to forward letters for the people; this is sufficient. Why should the government establish a post office to compete with the people? Moreover, it is doubtful whether the government can compete with them.

D: Anyhow, it is undignified for the government [to engage in trade] in opening a letter office. [22]

Such being the reaction of an educated group, it is small wonder that the Yamen ministers had procrastinated so long before adopting the postal reform project. The imperial decree, however, was received with general acclaim in the press, and the opening of the Imperial Post Office was awaited with enthusiasm. [23]

## Chapter VIII

## RETARDATION OF MODERNIZATION AND ITS CONSEQUENCES

The history of postal communication in China and the attempts at postal modernization between the years 1860-1896 reflects certain aspects of the contemporary Chinese society undergoing transformation in the process of contact with Western nations and some of the effects of those contacts. It shows in particular the sophistication and stability of the traditional institutions in China, the serious deficiency of knowledge and understanding of the literati about the world outside China, and the lack of strong incentive for institutional reform in mid-nineteenth-century China. Furthermore, this study examines the divers effects of China's contact with the West—the stimulation and disruption of native economic life and institutions, the introduction of Western "know-how" and assistance, and most important of all, the gradual awakening of the Chinese people to change and modernization.

The Ch'ing Official Post, particularly the I-chan, was considered by most Chinese to be the most efficient system for postal transmission and travel ever devised by man and perfected over centuries of experience. On the whole the I-chan of the nineteenth century met the requirements of a pre-modern society and continued to render important service in many areas for some time after the establishment of the Imperial Post Office. The same applies to the commercial letter agencies; they enjoyed the confidence of the people long after the Post Office offered a cheaper and sometimes a faster service. In spite of the stability of the Ch'ing Official Post, changing conditions after 1860 undermined its effectiveness. No provision was made for the I-chan or the P'u to handle new postal needs such as diplomatic correspondence and mail from abroad. The task was entrusted to organizations better equipped than the native institutions or to new institutions, namely, the Customs Post and the *wen-pao-chü*. These were merely expedients, however, for in doing so the government inadvertently reduced the importance of the traditional post, eventually neglecting it at a time when reform was most needed, when the coming of steamships and railroads called for the replacement of man and animal power to speed up the service. Many officials simply by-passed the I-chan, causing the service to deteriorate further. Western impact had just the opposite effect on the *hsin-chü*, which were flexible enough to utilize the new modes of transportation and flourished as a result of increased trade and general activity, economic or otherwise, in the treaty ports and the big cities. Ironically, the very stability and relative efficiency of the I-chan and the *hsin-chü* blinded the Chinese officials to the need for postal reform. Even after they had agreed to consider the postal project that Hart and others had presented, they were half-hearted and their attention was frequently distracted by seemingly more urgent matters. Hart reminisced that several Yamen ministers, governors-general, and Customs officials had had occasion to administer the postal project, but "differences of opinion at the capital and in the provinces and changes from time to time not only in the occupants of the official posts most concerned, but in the earlier and later views of those officials themselves, have combined to discourage effort and delay action until now." [1]

There were obviously many problems, one of which was the very nature of the modern postal service—a unified national postal system serving both the government and the people, handling both domestic and foreign mail in accordance with international postal rules. Since the I-chan and the P'u formed an integral part of the Ch'ing governmental structure, postal reform would affect local governments throughout the empire. Since the Western postal system was so different in principle and practice from the Chinese, an entirely new institution would have to be created. Unlike the railroad and the telegraph, the postal project could not be adopted piecemeal, and only the imperial government in Peking could handle it. The Yamen ministers were unwilling to assume the responsibility especially toward foreign powers until they were able to fulfill their obligations. They were afraid of opposition from vested interests which might vitiate their efforts entirely and they were reluctant to compete with the people who might lose their livelihood if the traditional systems were abolished;[2] there was also a shortage of funds and trained personnel. These were real and major problems, but if the ministers were determined to have a post office the obstacles were not insurmountable.

What were the consequences of the delay in creating a modern post office for China? In the first place, by postponing the establishment of monopoly on postal services the Chinese government gave the agencies the opportunity to grow and multiply, thus creating many problems for itself later when it decided to assume control.[3] For some years after 1896 the *hsin-chü* were allowed to carry on their business under special government regulations. They were not allowed to open branches in areas where the Imperial Post Office already existed, but by the latter years of the century they had become so powerful and numerous that they resisted control even though their days were numbered. It was only with great difficulty that the *hsin-chü* were finally suppressed in 1935.[4] Equally stiff competition came from the foreign post offices in China. By 1896 these establishments had doubled in number since 1861.[5] Foreign powers refused to give up their privileges and during the scramble for concessions, opened more post offices in China.[6] Only after much effort was the Chinese government able to secure a promise at the Washington Conference that these offices be withdrawn from Chinese soil by January 1923.[7]

The merits of the Post Office were recognized not long after its establishment, and several governors-general and governors urged Hart to extend its services to their provinces. By 1903 "there were almost a thousand post offices great and small scattered over the Empire and they were extending and expanding every month," as Hart proudly wrote a friend in England.[8] Just before the end of the dynasty the I-chan, which had by then outlived its usefulness, was abolished. In less than a quarter of a century after its inauguration the Post Office became a self-sustaining public service par excellence.[9] The serious regard for correspondence, official or private, and the tradition of efficiency and trustworthiness in postal transmission stood the Chinese staff well in the new institution. Yet, the successful metamorphosis from an ancient to a modern institution owed much to the experience gathered from the Customs Post.

The Chinese Post Office adopted the most up-to-date postal methods especially those of the British Post Office, modified to suit conditions in China. Since the Post Office was nurtured by the Customs, where the British civil service system had proved satisfactory, the latter was carried over to the postal service when it became independent of the Customs in 1911.[10] The Chinese Post Office became a model of efficiency and probity and was held up for other government institutions to emulate.

Without Sir Robert Hart's vision of progress for China and his persistence the establishment of a modern postal system might have been postponed for an even longer time. Even if a post office had been established, it might have had quite a different history, or it might have shared the fate of stunted growth of the steamship company and the railroads. Hart's unique position and his long tenure at the Maritime Customs, which provided him with competent personnel, facilities, and resources, enabled him to work out long range plans for the Post Office and lay a firm foundation for the infant institution before he retired.[11] He and the many Customs commissioners who worked in the Chinese civil service provided excellent examples of cooperation between East and West.

In retrospect, if postal communication in China had been modernized two or three decades before 1896, it could not have failed to help to promote trade and industry, facilitate the flow of knowledge and ideas, reduce regional barriers and affect people's attitude toward change, and in some ways hasten the much needed reform and modernization in Chinese society in general. The same may be said of other existing institutions. Unfortunately, in the absence of strong leadership at the helm of government, court intrigues and corruption became rampant. Those who wished to undertake reform were frequently frustrated and their initiative stifled. Modernization projects in the late Ch'ing period tended to be sporadic and fragmentary, poorly planned and hastily undertaken. The hopes built up during the period of national self-strengthening were shattered by the war of 1894-1895. The series of events which followed—internal conflicts and disturbances and foreign aggression—not only further impaired the power and prestige of the Ch'ing government but also increased its difficulties. The second wave of reform at the beginning of the present century came too late and the well-established modern institutions were too few in number to influence the final fate of the dynasty, even though in the last analysis much still depended on political factors.

106

## Abbreviations Used in the Notes

| | |
|---|---|
| *BPP* | *British Parliamentary Papers.* |
| *CHYK* | *Chung-ho yüeh-k'an.* |
| *CKHP* | *Huang ch'ao chang-ku hui-pien.* |
| *CTLT* | *Huang-ch'ao cheng-tien lei-ts'uan.* |
| *CTS:YCP* | *Chiao-t'ung shih: Yu-cheng pien.* |
| *HJAS* | *Harvard Journal of Asiatic Studies.* |
| *HT* | *Ta-Ch'ing hui-tien.* |
| *HTSL* | *Ta-Ch'ing hui-tien shih-li.* |
| *IWSM* | *Ch'ou-pan i-wu shih-mo.* |
| *NCH* | *North China Herald.* |
| *WCSL* | *Ch'ing-chi wai-chiao shih-liao.* |
| *WHTK* | *Ch'ing-ch'ao hsü wen-hsien t'ung-k'ao.* |

I. *The Problems of Modernizing Postal Communication in
Nineteenth-Century China*

1. For the Anglo-Chinese War, 1858-1860, see H.B. Morse, *International
Relations of the Chinese Empire* (New York and London: Longmans and
Green, 1910-1918), vol. 1, chaps. 16, 22-26. See provisions of treaties in
China in *Treaties, Conventions, etc. Between China and Foreign States*
(Shanghai: Inspectorate General of Customs, 1917), I, 404-421.

2. Teng Ssu-yü and John K. Fairbank, *China's Response to the West: A
Documentary Survey, 1839-1923* (Cambridge, Mass.: Harvard University
Press, 1954), pts. 3 and 4, passim; Mary C. Wright, *The Last Stand of Chinese
Conservatism: The t'ung-chih Restoration, 1862-1874* (Stanford: Stanford
University Press, 1957); *IWSM:TC*, 27:27, and chüan 41, 45, 53, and 54,
passim.

3. For the origin of the Inspectorate of Customs, see John K. Fairbank,
*Trade and Diplomacy on the China Coast: The Opening of the Treaty Ports,
1842-1854* (Cambridge, Mass.: Harvard University Press, 1953); Stanley F.
Wright, *Hart and the Chinese Customs* (Belfast: Wm. Mullen and Son for
Queen's University, Belfast, 1950), chaps. 4,5, and 8.

4. Inspector general's Circular No. 706 (Apr. 6, 1896) in *Documents
Illustrative of the Origin, Development, and Activities of the Chinese Customs
Service* (Shanghai: Inspectorate General of Customs, 1937), II, 42.

5. Howard Robinson, *Britain's Post Office* (Oxford: Oxford University
Press, 1953).

6. *BPP*, "Result of Inquiry with Abstract of Evidence," *Third Report of
the Select Committee on Postage*, XX, 12, 21-25 and passim.
To illustrate unsatisfactory postal services in England before 1840, one may
recall the familiar anecdote about Samuel T. Coleridge. He once paid the one
shilling postage (which he could hardly spare) for a woman who seemed unwilling to accept a letter from the carrier, thinking that she could not afford
the shilling. It turned out that the letter was simply a blank sheet of paper,
and the woman's son had sent it merely to inform her that he was well. On
another occasion, after paying two shillings for a letter in which the sender
lauded Coleridge as "immortal," Coleridge remarked, "I was ungrateful enough
to consider Mr. Hayden's immortality dear at two shillings . . . to this day my
thanks have not been given to Mr. Hayden for his apotheosis." See Samuel
Taylor Coleridge, *Letters, Conversations and Recollections* (London, 1836),
II, 114-115.

7. *BPP, Third Report*, p. vi.

8. Sir Rowland Hill, *Post Office Reform; Its Importance and Practicability*
(London, 1837); George Birbeck Hill and Sir Rowland Hill, *Life of Sir Rowland
Hill, K.C.B. and A History of the Penny Postage* (London, 1880).

9. The United States began to reform its postal service by reducing its rates in 1845; France reduced her postal rates in 1849. See A.D. Smith, *The Development of Rates of Postage* (London: Allen and Unwin, 1917), pp. 72-74; Alvin F. Harlow, *Old Post Bags* (New York and London: Appleton, 1928), pp. 332-334. In the United States, before the reduction in rates, the minimum postage for many years was six cents to send a letter within a radius of thirty miles. Thus a letter from Boston to Washington, D.C. (over 400 miles) would cost over twenty-five cents. Postal facilities did not keep up with settlements and expansion. For instance, until nearly 1800 there was not a single post office on Long Island. As for the Midwest, "any one who made a trip east from a frontier village such as Chicago or St. Louis was apt to have his bags so full of other people's mail that he scarcely had room for his tooth-brush." Private companies were formed to carry the mail more quickly and cheaply than the official postal service, but until the first Pony Express service was inaugurated in 1825, there was no substantial improvement. See Harlow, pp. 283, 304, 321-327, 332; Smith, p. 71.

10. Howard Robinson, *The British Post Office: A History* (Princeton: Princeton University Press, 1948).

11. Much of the following information is derived from travel diaries mentioned later in this work, but I am also indebted to the article, "Inland Communications in China," ed. T.W. Kingsmill, *Journal of the China Branch of the Royal Asiatic Society,* 28:1-128 (1893-1894).

12. T'ao Pao-lien, *Hsin-mou shih-hsing chi* (Peking, 1896), 1:9.

13. William Soothill, *Timothy Richard of China* (London: Seeley Service, 1924), p. 48.

14. John K. Fairbank and Teng Ssu-yü, "On the Transmission of Ch'ing Documents," *HJAS*, 4.1:44-46 (May 1939).

15. For postal services in the West, see Harlow; J.C. Hemmeon, *The History of the British Post Office* (Cambridge, Mass.: Harvard University Press, 1912); Robinson, *The British Post Office* or *Britain's Post Office.*

## II. *The Ch'ing Official Post*

1. *Shuo-wen chieh-tzu ku-lin,* comp. Ting Fu-pao (Shanghai, 1937), V, 2792-93.

2. "Te-chih liu-hsing su-yü chih-yu erh ch'uan-ming," in Mencius, "Kung-sun Ch'ou," pt. 1. *Chih* and *yu* were translated as "postal transmission by mounted courier" and "postal transmission by foot courier" respectively in *Shuo-wen chieh-tzu ku-lin* VII, 3394; X, 4339. See *Tso-chuan,* passim.

3. For the traditional postal system, see Lou Tsu-i, *Chung-kuo yu-i fa-ta-shih* (Kunming: Chung-hua Press, 1940); Lou Tsu-i, *Chung-kuo yu-i shih-liao*

(Peking: Jen-min yu-tien chü-pan-she, 1958); Chang Liang-jen, "Chung-kuo li-tai yu-chih kai-yao," *Tung-fang tsa-chih*, 32.1 (*yen*): 7-16 (January 1935).

4. "Ching-chi pien, yung-cheng tien, i-ti k'ao," *Ku-chin t'u-shu chi-ch'eng*, ed. Chiang T'ing-hsi et al. (Shanghai, 1884), 258a:1-6.

5. Mencius, "Kung-sun Ch'ou," pt. 1.

6. "Wei Ku mou fa Yung," *Tso-chuan*, chüan 2; "Yu Chi shih-Ch'u," ibid., chüan 4; "Chien Shu chien hsi Cheng," ibid., chüan 2.

7. *T'ang liu-tien*, ed. Li Lin-fu et al. (Shanghai, 1895), 1:31, 3:41b, 5:32b-33b.

8. "Yung-ch'eng tien," *T'u-shu chi-ch'eng*, 258.1:7bff.

9. Edward Schafer, *The Golden Peaches of Samarkand: A Story of T'ang Exotic* (Berkeley: University of California Press, 1963), p. 33.

10. There are five editions of the *Ta-Ch'ing hui-tien:* 1690, 1732, 1748, 1818, 1899. For this study only the Chia-ch'ing edition (1818) and the Kuang-hsü edition (1899) were consulted. See *HT:CC*, chüan 39, *HT:KH*, chüan 51. See section on *yu-cheng* in *HTSL*, Kuang-hsü ed., chüan 655-703.

11. *HT:KH*, 51:3; *HTSL*, chüan 688-689.

12. *HTSL*, chüan 659-683.

13. *HT:KH*, 51:1a-b.

14. *HTSL*, chüan 655-658.

15. *HT:KH*, chüan 51. On the administration and the organization of the postal system, I am much indebted to Professors John K. Fairbank and Teng Ssu-yü for their work entitled "On the Transmission of Ch'ing Documents," *HJAS*, 4.1:12-46 (May 1939), and "Types and Uses of Ch'ing Documents," *HJAS*, 5.1:1-71 (January 1940).

16. For a different translation and interpretation, see H.S. Brunnert and V.V. Hagelstrom, *Present Day Political Organization of China*, tr. A. Beltchenko and E.E. Moran (Shanghai: Kelly and Walsh, 1912), p. 142; Fairbank and Teng, "Transmission," p. 37.

17. *HT:KH*, 51:17.

18. *HT:KH*, 51:16b.

19. For administrative purposes the province in Ch'ing times was divided into prefectures (*fu*), which were subdivided into districts (*hsien*). In some cases the prefecture also controlled a few sub-prefectures, which were often larger in size than the districts, but independent sub-prefectures (*chih-li-chou*) and independent departments (*chih-li t'ing*) were the same as prefectures and might also have districts under their jurisdiction.

20. *HT:KH*, chüan 52.

21. *HT:KH*, 51:1-2.

22. *Tung-hua hsü-lu*, ed. Chou Shou-p'eng (Shanghai, 1909), Kuang-hsü 18.2, 147:18.

23. Hsieh Pao-chao, *The Government of China, 1644-1911* (Baltimore: Johns Hopkins University Press, 1925), pp. 263-265; *HT:KH*, 51:5b-6.

24. Ibid.

25. Ibid.

26. *HTSL*, chüan 659.

27. *HTSL*, chüan 659-683.

28. *Ta-Ch'ing lü-li* (1870), chüan 22.

29. Marco Polo, *The Travels of Marco Polo the Venetian*, tr. William Marsden (London: J.M. Dent, 1923), bk. 2, chap. 20, p. 210.

30. *Ta-Ch'ing lü-li*, chüan 22.

31. *HTSL*, 701:4, 8b-11.

32. *HT:KH*, 51:12b-13b, 69; 13b-15.

33. Brunnert and Hagelstrom, p. 341.

34. *HT:KH*, 51:14.

35. *HT:KH*, 51:3, 6-10b; *HTSL*, 702:10.

36. *HT:KH*, 51:10; Fairbank and Teng, "Transmission," pp. 24-28.

37. *HTSL*, 684:5; *Ta-Ch'ing lü-li*, chüan 22.

38. *Liu (li)-pu ch'u-fen tse-li* (1887), chüan 35.

39. *HTSL*, 700:15-16.

40. *HTSL*, 700:16b. The waybills which were attached to some documents were lost and investigation was made about stations in Shensi, Shansi, and Chihli.

41. Tseng to his brothers, Hsien-feng 10.6.28 (Aug. 14, 1860). In the thick of battle dispatches were delayed as much as five days; Tseng to another brother, Hsien-feng 10.12.4 (Jan. 14, 1861). See Tseng Kuo-fan, *Tseng Wen-cheng-kung chia-shu* (Shanghai: Commercial Press, 1905), 7:1b, 3b.

42. Memorial of Hu Lin-i (1852) in *WHTK*, 375:16.

43. Ibid.; *HTSL*, 697:11b.

44. *WHTK*, 375:16.

45. *HTSL*, 697:13.

46. *Mu-tsung I-huang-ti sheng-hsün*, in *Shih-ch'ao sheng-hsün* (Peking, n.d.), 138:9bff; *HTSL*, 702:26b-27; *WHTK*, 375:16.

47. *HTSL*, 702:26b-27.

48. *Tung-hua hsü-lu*, 123:5.

49. The "express warrant" specified as follows:

"_____(Government office or surname and official title of sender), having an urgent document to be delivered to_____(name of official) at _____ yamen, requests that postal stations enroute furnish without distinction of day or night, men and horses and forward the said document at the rate of_____li per day, posthaste. Any damage noticed on the document should be reported to the official from whom the document originated."

See "Ch'ing-tai chou-hsien ku-shih," ed. Ts'ai Shen-chih, in *CHYK*, 2.12: 107-108 (December 1941).

50. *HT:KH*, 51:10b-11b.

51. Ibid.

52. Fairbank and Teng, "Transmission," p. 37.

53. *IWSM:TC*, 96:25b-27.

54. *IWSM:TC*, 99:1.

55. *Tung-hua hsü-lu*, 109:1.

56. Fairbank and Teng, "Transmission," p. 46.

57. Ibid., pp. 41-45.

58. *HTSL*, chüan 693.

59. *HT:KH*, 51:6.

60. Ibid.; *HTSL*, chüan 698-699; *Ping-pu chung-shu cheng-k'ao* (Peking, 1825), chüan 34.

61. Ibid., 36:13-18; *HTSL*, chüan 684.

62. Shen Kuei-fen, "Shen wen-ting Yüeh-yao jih-chi," *CHYK*, 1.4:95-103 (April 1940), 1.5:79-84 (May 1940), 1.6:92-98 (June 1940).

63. Ch'ü Hung-chi, *Shih-Yü, shih-Min jih-chi*, ed. Ch'ü Hsüan-ying (Peiping, 1933).

64. T'ao Pao-lien, passim.

65. Polo, bk. 2, chap. 20, p. 208.

66. Weng T'ung-ho, *Weng Wen-kung-kung jih-chi* (Shanghai: Commercial Press, (1925), Hsien-feng 8.7.6 (Aug. 14, 1858), 1:1; Ch'ü Hung-chi, pp. 2, 2b, 6b, 9, 12, 13; T'ao Pao-lien, chüan 3-6 for accommodations in the north-west.

67. Shen Ping-yüan, *Shen Wen-chieh-kung hsing-yao jih-chi* (Szechwan, 1885), p. 5.

68. Shen Ping-yüan mentioned that he was allowed nine horses by the Ping-pu, and he hired at his own expense three pack-horses (or mules) and six men: a guide, two men to look after the baggage, two to follow the sedan-chairs, and a head chair-bearer. See Shen Ping-yüan, p. 5. When Shen Kuei-fen crossed the Yangtze at Shasi he used between twenty and thirty boats. *CHYK*, 1.4:99. In 1875 another examiner, Ch'ü Hung-chi hired seven carts from the Honan-Chihli borders all the way to Peking. The rental for one carriage-like cart was 3,500 copper cash, and for the other six (for servants and baggage) was 42,000 cash, approximately thirty silver dollars. See Ch'ü Hung-chi, p. 5.

69. Shen Kuei-fen, Hsien-feng 11.7.29, 11.7.30 (Sept. 3 and 4, 1861), in *CHYK*, 1.4:98 (April 1940).

70. Ibid.

71. Ibid.

72. "Ko-hang shih-chien," reprint in "Ch'ing-tai chou-hsien ku-shih," *CHYK*, 2.10:99-101 (October 1941), 2.11:100-101 (November 1941), 2.12:100 (December 1941); "Kung-men yao-lüeh," *CHYK*, 2.11:101.

73. Ch'ü T'ung-tsu, *Local Government in China Under the Ch'ing* (Cambridge, Mass.: Harvard University Press, 1962), chap. 5.

74. Ibid., pp. 61-62.

75. Some 200 members of a Liu clan who had served as runners in a district yamen in ch'ang-yih (near Lai-choufu), Shantung were dismissed from government service by order of the governor, and their leader was decapitated. They were convicted for having embezzled several hundred taels in handling supplies to troops, for asking exorbitant prices for carts and men, and for intimidating and extorting money and vehicles from the country people contrary to the magistrate's commands. An imperial decree approving the Shantung governor's action asked other governors-general and governors to ferret out similar ruffians elsewhere. *Peking Gazette* (June 28, 1895), pp. 83, 86-87. Cf. Ch'ü T'ung-tsu, chap. 4.

76. Ibid., pp. 87-88.

77. Ibid., p. 88.

78. Shen Kuei-fen, Hsien-feng 11.12.15 (Jan. 14, 1862), *CHYK*, 1.6:93.

79. See various diaries, for example, Shen Kuei-fen, Hsien-feng 11.11.24 (Dec. 25, 1861), *CHYK*, 1.5:79, 1.6:93, 95: Ch'ü Hung-chi, Kuang-hsü 17.81 (Sept. 3, 1891) p. 23; T'ao Pao-lien, Kuang-hsü 17.5.27, 17.7.23, 17.8.7, 17.10.2, 17.12.7 (July 3, Aug. 27, Sept. 8, Nov. 3, 1891; Jan. 6, 1892), 1:33, 1:43, 2:8b, 3:48b, 6:49.

80. Shen Kuei-fen remarked that magistrates in places off the beaten track were more spontaneous and warmer in their reception. See T'ung-chih 1.2.7 (Mar. 7, 1862), *CHYK*, 1.6:96.

81. T'ao Pao-lien, Kuang-hsü 17.9.13 (Oct. 15, 1891), 3:3.

82. *Chang Wen-hsiang-kung ch'üan chi* (Peiping, 1928), *kung-tu*, 86:8.

83. *Huang-ch'ao ching-shih-wen, hsü-pien,* comp. Ko Shih-chün (Shanghai, 1896), 67:6-7.

84. See article by Ting Pao-chen, ibid., 67:7.

85. *Chang Wen-hsiang-kung ch'üan-chi, kung-tu,* 86:31b-34.

86. Ibid., 86:8b, 33b.

87. Ibid., 86:32b-33.

88. Arthur W. Hummel, ed., *Eminent Chinese in the Ch'ing Period, 1644-1912* (Washington, D.C.: Library of Congress, 1943-1944), I, 28.

89. Ch'ü T'ung-tsu, p. 3.

90. Liu Yüeh-yün, comp., *Kuang-hsü k'uai-chi piao* (N.p., 1901), 1:3-4b.

91. *Honan-fu-i ch'üan-chi* (Honan, 1893), 1:139b-145; *Honan ts'ai-cheng shuo-ming-shu* Honan, 1909), supplement, 1:1b.

92. Liu Yüeh-yün, 1:3-4b.

93. *HT KH,* 51:2.

94. Shen Kuei-fen, *CHYK* 1.6:93.

95. The figure of 3 million taels was mentioned in many memorials and essays. Some of these were quoting an essay attributed to Feng Kuei-fen (1828-1874), "On the Abolition of the I-chan" (T'sai i-chan i). However, the essay was not found in Feng's collected works, and certain anachronisms in the essay makes his authorship doubtful. This essay was widely known and printed in many collections such as the *Ching-shih wen, Hsü Wen-hsien t'ung-k'ao,* and others. Chang Chih-tung and Liu K'un-i in their joint memorial for reforms in 1901 also used the figure "three millions" and the sum was repeated by Théophile Piry in "Report on the working of the Post Office of 1904," *Decennial Reports, 1892-1901* (Shanghai: Inspectorate General of Customs, 1906), II, p. 314. See also, *CTLT*, 451:9b-10b, 463:7, and *WHTK*, 376:16. On the other hand, the modern historian Hsiao I-shan believed that I-chan expenses amounted to 2 million taels per year. See *Ch'ing-tai t'ung-shih* (Shanghai, 1922), 2:397.

## III. *The Letter Agencies*

1. The difficulty of communicating by post has been well documented in biographies, diaries, and other literature. Shen Kuei-fen, who went to Canton in 1861 as an imperial examiner, sent letters home every week or ten days by sundry means. When he presented three memorials (one to the emperor, two to the dowager empresses) after his new appointment as senior vice-president of the Board of Rites, he entrusted them to the mounted couriers of the governor of Kwangtung and paid them thirty dollars. In addition, he asked them to take a letter to his family in Peking--he had never sent private mail by official couriers during his trip--in which he instructed his family to give the couriers four taels as tea money. See Shen Kuei-fen, *CHYK*, vol. 1, nos. 4, 5, and 6, passim; also, 1.4:96-99, 1.6:93, 97. See especially entries for Hsien-feng 11.12.15 (Jan. 14, 1862), T'ung-chih 1.2.17 (Mar. 17, 1862).

2. Yin Hsien (Hung-ch'iao), a magistrate in the Chin dynasty (A.D. 265-419), was said to have carried at one time over a hundred letters as a favor for people in the capital. When he arrived at a place called Shih-t'ou, he threw all the letters in the stream and said: "Those that sink will sink by themselves and those that float will do likewise. I, Yin Hung-ch'iao, am not going to be a postman," See "Yin Hsien," *T'zu-yüan* (Shanghai, 1948), *ch'en,* p. 124. It seems that in the fifth century there were already individuals who carried letters for a living, and Yin Hsien was annoyed for being so burdened for nothing.

3. Piry, *Decennial Reports, 1892-1901,* II, xlvii; I, 509. Also, H.B. Morse, *Trade and Administration of the Chinese Empire* (Shanghai: Kelly and Walsh, 1908), p. 379. Tōa Dobunkai, comp., *Shina keizai zensho* (Tokyo, 1908),

VI, 100, places the beginning of the agencies in the reign of T'ung-chih (1862-1874), but some postal firms were known to have been established long before that time; for example, Hu Wan-ch'ang in Chungking was founded in 1822. See *Decennial Reports, 1892-1901*, I, 174-175.

4. Hsieh Pin, *Chung-kuo yu-tien hang-k'ung-shih* (Shanghai:Chung-hua Press, 1946), p. 19. According to various reports of the Chinese Maritime Customs, the *hsin-chü* continued to thrive in the eighties and nineties.

5. "Shanghai," *Decennial Reports, 1892-1901*, I, 509.

6. Some seven agencies in Hankow, five in Chefoo, ten in Wuhu, and about half a dozen in Amoy and Swatow respectively all had their head offices in Shanghai while the half dozen or so "steamer letter agencies" in Canton had representatives and agents in all treaty ports. See *Decennial Reports, 1882-1891* (Shanghai: Inspectorate General of Customs, 1893), under reports of various ports just mentioned.

7. "Shu-ch'ien-chia," in Hsieh Pin, p. 19.

8. Piry, *Decennial Reports, 1892-1901*, II, xlvii.

9. An article in *T'oung pao* estimated that there were 200 postal agencies in Shanghai; no reference was given as to how the figure was obtained. See "Postal Service in China," *T'oung pao*, 5:64 (1896).

10. "Tientsin," *Decennial Reports, 1892-1901*, II, 570.

11. *Decennial Reports, 1882-1891*, p. 380; *Decennial Reports, 1892-1901*, I, 459.

12. Ibid., p. 117.

13. A distinguished librarian of a famous university.

14. *Ch'i Ju-shan sui-pi* (Taipei: Chung-yang wen-wu kung-ying-she, 1953), pp. 69-73.

15. *Decennial Reports, 1882-1891*, p. 226; *Decennial Reports, 1892-1901*, I, 358-359.

16. These were the rates at Yochow, Hunan, but postage in other places was about the same.

17. Payments were made in part or in full, depending on the circumstances. For example, if a courier was killed in an accident or robbed, no compensation would be made.

18. In Amoy, owing to the hilly conditions, a messenger would only be able to cover approximately fifty li a day. The charge for a three-day journey was 2,000 cash, for a six-day journey, 6,000 cash. See *Decennial Reports, 1882-1891*, p. 520.

19. Ibid., pp. 73, 155.

20. Ibid., p. 155.

21. Ibid., p. 535.

22. Ibid.

23. Morse, *Trade and Administration*, p. 379.

24. "Chungking," *Decennial Reports, 1882-1891*, p. 116. (An aggregate of 200 li in twenty-four hours.)

25. *T'oung pao,* 5:64 (1896).

26. "Chungking," *Decennial Reports, 1882-1891*, p. 117.

27. Piry, *Decennial Reports, 1892-1901*, II, xlvii.

28. "Kiukiang," *Decennial Reports, 1882-1891*, pp. 224-227.

29. The number of *hsin-chü* rose from sixteen in 1897 to nineteen in 1901. See *Decennial Reports, 1892-1901*, II, 358.

30. *Decennial Reports, 1882-1891*, p. 226.

31. Ibid., pp. 226-227.

32. In Chungking sixteen postal agencies worked together; three were engaged in postal transmission between Chungking and Hankow, the others in postal work in the interior. See ibid., p. 117.

33. Ibid., pp. 404, 535.

34. Cecil A.V. Bowra, son of Senior Commissioner E.C. Bowra, at Newchang See *Decennial Reports, 1892-1901*, I, 30.

35. "Ichang," *Decennial Reports, 1882-1891*, pp. 154-155.

36. These agencies often had large amounts of silver to forward at regular intervals either for the government or for commercial firms, and they were staffed with men particularly skilled in Chinese pugilism and the use of arms who usually had a highly developed sense of honor and chivalry. The escort agencies sometimes had a special understanding with bandit gangs so that their convoys could pass the latters's "territory" unmolested. They also carried letters for their clients. See Ch'i Ju-shan, "Man-t'an Chung-kuo yu-shih" and "Piao-chu-tzu shih-hua" in *Ch'i Ju-shan sui-pi*, pp. 69-73.

37. *Decennial Reports, 1892-1901*, II, 141.

38. Yang Lien-sheng, *Money and Credit in China* (Cambridge, Mass.: Harvard University Press, 1952), p. 81.

39. Ch'en Ch'i-t'ien, *Shan-hsi p'iao-chuang k'ao-lüeh* (Shanghai: Commercial Press, 1937), pp. 34-35, 69-78, 97-98. Natives of Shansi dominated the remittance bank business as natives of Ningpo dominated the letter-carrying business. In Shansi alone there were fort  ine Shansi banks and eleven other banks; the former had 414 branches sca  :ed in eighty-three locations, the larger ones having connections in as many as thirty places.

40. Ch'en Ch'i-tien, p. 110. Shanghai, *Decennial Reports, 1892-1901*, I, 508.

41. Ch'en Ch'i-t'ien, pp. 161-162.

42. *Decennial Reports, 1892-1901*, II, 143. Archibald R. Coloquhoun, *China in Transformation* (New York, 1891), p. 108.

43. *Decennial Reports, 1882-1891*, p. 535; *Decennial Reports, 1892-1901*, II, 142.

44. Ko Kung-chen, *Chung-kuo pao-hsüeh-shih* (Shanghai: Commercial Press, 1927), p. 23 ff., especially p. 28. For superintendents of posts, see *HT*, 51:16b; also, F.W. Mayers, "*The Peking Gazette*," *China Review*, 3.1:13-18 (July and August 1894).

45. Ko kung-chen, p. 35.

46. Ibid., p. 41; *Ch'i Ju-shan sui-pi*, p. 33.

47. Ko Kung-chen, p. 76.

48. Among those published under Western auspices: *Hua-tzu jih-pao*, published by the *Daily Press* (Hong Kong, 1864-1865); *Shang-hai hsin-pao*, published by the *North China Daily News* (Shanghai, 1862), and *Shen-pao*, published by an Englishman named F. Major (Shanghai, 1872). See ibid., p. 76 ff. Among those published by Chinese proprietors were : *Chao-wen-hsin-pao* (Hankow, 1873); *Hui-pao*, organized by Yung Wing (Shanghai, 1874); *Hsün-huan jih-pao*, edited by Wang T'ao (Hong Kong, 1874). See ibid., pp. 115, 121-122.

49. Yao Kung-hao "Shang-hai pao-chih hsiao-shih," *Shang-hai hsien-hua* (Shanghai: Commercial Press, 1926), 2:107-109.

50. "*Shen-pao* kuan t'iao-li," in Ko Kung-chen, p. 83.

51. "Chinkiang," *Decennial Reports, 1892-1901*, I, 459; "Tientsin," ibid., II, 570.

52. Yao Kung-hao, passim.

53. Report by Commissioner F. Hirth in *Decennial Reports, 1882-1891*, p. 315.

54. In the Canton delta there were individual couriers as late as the 1920s who traveled regularly between Canton, Macao, Hong Kong, and the surrounding districts, visiting each place at definite intervals. They were known as *hsün-ch'eng ma* (lit., "the horse that tours the city"). In many places, however, such facilities were lacking. For instance, Mengtsz was opened as a port in 1889, but as late as 1891 it had only the government post. The Maritime Customs had its newspapers and private letters forwarded from Canton by a bank in Yunnanfu. The mail often came in a roundabout way via Shanghai, Hankow, and Chungking. Most private correspondence was sent from Mengtsz by muleteers to Yunnanfu, where there were two *hsin-chü* at the time.

## IV. *Foreign Postal Establishments in China*

1. Morse, *International Relations*, 1, 72.

2. Ibid., p. 344.

3. *CTS: YCP*, III, 1,307.

4. Morse, *International Relations*, I, 82, 89-90.

5. G.T. Bishops, C.S. Morton, and W. Sayers, *Hong Kong and the Treaty*

*Ports: Postal History and Postal Markings,* 2nd ed.; rev. and enlarged ed., Harold E. Lobdell and Adrian E. Hopkins, (London: The Postal History Society, 1949), p. 18. The first part of this revised work for which Mr. Lobdell is chiefly responsible is probably the most comprehensive single account of the Hong Kong Post Office and the British packet-agencies as well as the origin of other postal agencies in China. I am greatly indebted to this work for much of materials in this chapter.

6. Ibid., p. 35.

7. Morse, *International Relations,* I, 287-288.

8. Bishop, Morton and Sayers, pp. 36-37, 39; Morse, *International Relations,* I, 288.

9. Ibid., p. 344.

10. Ibid.

11. Bishop, Morton and Sayers, pp. 41, 43.

12. James Legge, "The Colony of Hongkong: A Lecture on Reminiscences of A Long Residence in the East," *The China Review,* 1.3:163-176 (November-December 1872).

13. Bishop, Morton and Sayers, pp. 43, 47-48.

14. The frigate *Horatio,* which left England early in July 1815, did not reach Canton until mid-January 1816 and the supercargoes of the East India Company did not hear about the victory at Waterloo until then. See H.B. Morse, *The Chronicles of the East India Company Trading in China, 1635-1834* (Oxford: Oxford University Press, 1926), IV, 229-230. For the search and adoption of the overland route see *First Report of the Postmaster General on the Post Office* (London, 1855), p. 18: Harlow, pp. 196-199; Geoffrey Clarke, *The Post Office of India and Its Story* (London: John Lace and Bodley Head, 1921), pp. 122-125.

15. Fairbank, *Trade and Diplomacy,* pp. 168-189; Morse, *International Relations,* I, 345.

16. Ibid.

17. Fairbank, *Trade and Diplomacy,* p. 168; Bishop, Morton and Sayers, pp. 50, 58, 61.

18. Legge, p. 166. The passengers spent fifty-five days on board on their way from London. With different connections, the mail took only fifty days. See *Chinese Repository,* 14.8:400 (August 1845); Fairbank, *Trade and Diplomacy,* pp. 168-169.

19. Bishop, Morton and Sayers, pp. 64-65.

20. Ibid., p. 50.

21. *NCH* (Aug. 4 and Dec. 29, 1860). Days and months of arrivals in Shanghai between 1850 and 1860 were listed under the heading "Arrival of Mail." Bi-monthly arrivals began in April 1853. See Morse, *International Relations,* I, 345.

22. *NCH* (Aug. 11, 25, Sept. 11, 22, 29, and Oct. 20, 1860; Feb. 2 and Mar. 16, 23, 1861).

23. Fairbank, *Trade and Diplomacy,* p. 171.

24. Morse, *International Relations,* I, 345-346.

25. Bishop, Morton and Sayers, p. 75. An agency opened at Whampoa in 1858 or 1859 was closed permanently in 1863.

26. Ibid., pp. 89-90.

27. *Fourteenth Report of the Postmaster General on the Post Office* (London, 1868), p. 14.

28. *NCH* (Feb. 9, 1861).

29. Bishop, Morton and Sayers, pp. 93-94.

30. Ibid., p. 87.

31. Ibid., p. 50.

32. Ibid., p. 101ff. *Annual Report of the Postmaster General of the United States* (Washington, D.C., 1867), p. 21; reprint of a description of the ship *Colorado* (3,700 tons) from *Japan Times,* in *NCH* (Feb. 8, 1867).

33. A convention between the United States and the Hong Kong post offices in 1867 agreed on rates between Hong Kong or Shanghai and the United States; eight cents for a letter weighing one-half ounce, and two cents for newspapers. The postage between the United States and Britain was six cents or three pence. Letters between Hong Kong and England cost forty-six or fifty-four cents via Southampton or Marseilles respectively, by the Suez route. Bishop, Morton and Sayers, pp. 105-106. The British mail did not use the Suez Canal (completed in 1869) until 1888 when the British government acquired a larger part of its shares.

34. See *Annual Report of the Postmaster General of the United States* for 1868, 1869, 1870, 1872.

35. Bishop, Morton and Sayers, p. 109.

36. Ibid., p. 122; Max Andersch, *Die Deutsche Post in der Turkei, in China, und in Morokko* (Berlin: R.V. Decker's Verlag, 1912), p. 100.

37. *Annual Report of the Postmaster General of the United States* for 1867 and 1868; Bishop, Morton and Sayers, pp. 77, 107, 122. During the campaign of 1858 France had a field post office in Shanghai; in 1862 it was turned into a civil post office. See Bishop, Morton and Sayers, p. 77. The Germany agency established in 1886 was called the Kaiserlich Deutsches Postamt until 1896. See Andersch, p. 100.

38. *Decennial Reports, 1892-1901,* II, 571.

39. Morse, *International Relations,* I, 348ff.

40. Bishop, Morton and Sayers, pp. 93-94.

41. Sir Robert Hart to Henry Kopsch, Dec. 15, 1892, and June 28, 1893. Mss. in Houghton Library, Harvard University.

42. S. Wright, pp. 316, 629-630; Bishop, Morton and Sayers, p. 124.

43. References to the new local posts appeared in *NCH* in Chefoo (July 27, 1894), in Chinkiang (Aug. 17, 1894), in Ichang (Dec. 28, 1894).

44. *NCH* (July 13, 1894; Aug. 24, 1894).

45. *NCH* (May 25, 1894); Bishop, Morton and Sayers, p. 125.

46. The experience of some French Jesuits at a small village in Ho-chien fu in southern Chihli is representative. Before the Chinese Post Office was established the Jesuits used to send a courier to Tientsin once every fortnight for letters, parcels, and newspapers which arrived by every French mail. Numerous steamers traveling between Shanghai and Tientsin during the summer would carry their letters free of charge. In Shanghai the Jesuits redistributed and redirected the mail. Letters to be sent to France were made into a bundle and sent free of charge with the consular post bag to the Ministry of Foreign Affairs in Paris, where the Father Procurator of the Jesuit order took over and forwarded them to their respective destinations. Thus the Jesuits paid about ten centimes less per letter than they would have had to pay to send letters directly from Ho-chienfu. The couriers took six days to make a round trip between Shanghai and Tientsin and charged 2,400 cash. See letter from Father Em. Becker to Father Hamann in Rheims, Aug. 28, 1899, in "Lettres des Missionaires de la Compagnie de Jesus, Province de Champagne," *Chine et Ceylan,* 1.4:298-299 (1898-1900).

47. The frequency of the service was left out in the Chinese text. See Article 11 of the Treaty of Tientsin, *Treaties, Conventions, etc.,* I, 90, 99.

48. Gaston Cahen, *Histoire des Relations de la Russie avec la Chine sous Pierre le Grand, 1689-1730* (Paris: Lib. Felix Alcan, 1912), pp. 221-223, 225.

49. Memorial of the Tsungli Yamen, *IWSM:HF,* 68:13b.

50. *Treaties, Conventions, etc.,* I, 110.

51. Godfrey E. Hertslet, ed., *China Treaties* (London: Her Majesty's Stationery Office, 1908), doc. no. 81, I, 461.

52. *IWSM:HF,* 75:21-23.

53. *IWSM:HF,* 76:7, 9, 10, 77:3-4.

54. Ibid., 76:29b, 77:3.

55. Ibid., 76:7b, 77:3. For circumstances in which the treaty was negotiated, see Ch'en Fu-kuang, *Yu-Ch'ing i-tai chih Chung-O kuan-hsi* (Yunnan, 1947), pp. 133-137.

56. I wish to thank Professor Benjamin I. Schwartz of Harvard University for his kindness in comparing the Chinese and Russian texts of the treaty for me.

57. *IWSM:HF,* 78:9 ff. In a convention for land trade concluded on March 4, 1862 (T'ung-chih 1.2.4), Russian traders who came by land to Tientsin were given a reduction on the regular Customs duties charged on those who came by sea. See *Chung-O yüeh-chang hui-yao,* comp. Tsungli Yamen (Peking, 1882), 3:9 ff, 13 ff. Cf. Hertslet, ed., I, 478 ff.

58. Li Sung-p'ing, *K'e-yu wai-shih* (Hong Kong: Pao-an yu-piao-she, 1966), pp. 108-111; Bishop, Morton and Sayers, p. 172.

59. *Treaties, Conventions, etc.,* I, 110.

60. *IWSM:HF,* 80:35b.

61. Ibid., pp. 34-35b.

62. Ibid., 75:24b, 76:10, 78:25b.

63. Ibid., 79:12.

64. Cf. ibid. and ibid., 75:22 ff.

## V. *Origin and Development of the Customs Post*

1. *Documents,* V, 375-388.

2. S. Wright, pp. 196, 198.

3. Inspector General's Circular No. 706, Peking (Apr. 9, 1896) in *Documents,* II, 42.

4. S. Wright, p. 316.

5. For provisions of the Treaties of Tientsin, see *Treaties, Conventions, etc.,* I, 404-411.

6. Ch'en Tse-chuan, *Chung-Kuo tsao-ch'i yu-cheng ti shih-liao k'ao-shih* (Taipei: Cathay Philatelic Library, 1965), p. 6.

7. Treaties with Germany in 1861, Portugal in 1862, Denmark in 1863, Spain in 1864, Belgium in 1865, Italy in 1866, Austria-Hungary in 1869, and Japan in 1871. See *Treaties, Conventions, etc.,* I, 404 ff. See British treaty, ibid., p. 406; French treaty, ibid., p. 815; Hertslet, ed., doc. no. 20, I, 270.

8. *Treaties, Conventions, etc.,* I, 406; Hertslet, ed., doc. no. 6, I, 20.

9. Bishop, Morton, and Sayers, pp. 165-174.

10. See *IWSM:HF,* 71:19 ff., *IWSM:TC,* 28:13-17; for Yamen's organization see *HT,* 99:1-2b, 10; *HTSL,* chüan 1220.

11. For the creation of the Inspectorate of Customs, see Fairbank, *Trade and Diplomacy*, pt. 5; S. Wright, chaps. 4,5,8.

12. Morse, *International Relations,* III, 62; S. Wright, pp. 316-317.

13. Ibid.

14. *IWSM:HF,* 72:20b; *HTSL,* 1220:2.

15. Piry, *Decennial Reports, 1892-1901,* II, xlv.

16. "Chinkiang," *Decennial Reports, 1882-1891,* p. 314.

17. Juliet Bredon, *Sir Robert Hart: The Romance of a Great Career* (London: Hutchinson, 1909), pp. 158-159.

18. Peiho froze sometimes between the months of November and March. *Customs Gazette* (January-March, 1869); ibid (January-March 1871), p. 35.

19. Ibid. (January-March 1870), p. 30.

20. Piry, *Decennial Reports, 1892-1901,* II, xlv.

21. *Report of the Chinese Post Office for the Tenth Year of Chung-hua Min-kuo (1921), with which is Incorporated an Historical Survey of the Quarter-century, 1896-1921* Shanghai: Ministry of Communication, Directorate General of Posts, 1922), appendix B.

22. Ibid.

23. Ibid., p. 34.

24. Wu Hsiang-hsiang, *Wan-Ch'ing kung-t'ing shih-chi* (Taipei: Cheng-chung shu-chü, 1952), chap. 2, pp. 95-151; chaps. 1-4, passim.

25. *IWSM:TC,* 86:27b.

26. *IWSM:TC,* 27:27; Alexander Michie, *The Englishman in China during the Victorian Era: Life of Sir Rutherford Alcock* (London, 1900), II, 136.

27. *IWSM:TC,* 40:10b-23, esp. p. 20.

28. Ibid., 40:10b ff, 23-37. The memorandum was translated by H. E. Woodhouse and printed many years later as "Mr. Wade on China," in *China Review,* 1.1:38-44 (July-August 1872); 1.2:118-123 (September-October 1872).

29. *IWSM:TC,* chüan 41, 45, 53, 54, passim.

30. Teng and Fairbank, pt. 3, pp. 61-83; M. Wright, chaps. 9, 10.

31. *IWSM:TC,* 47:24-25; 48:1-5, 10b-15, 18; 49:14-18b, 23b-24.

32. Wu Hsiang-hsiang, pp. 120-124, 126-127.

33. Wen-hsiang, *Wen Wen-chung-kung shih-lüeh* (N. p., 1882), 1:12b.

34. "Title of I-cheng wang" conferred. See Wu Hsiang-hsiang, pp. 99-114.

35. The Summer Palace (Yuan-ming yuan), burnt by the allies in 1860, was to be partially rebuilt, ostensibly for the pleasure of the empresses dowager in their retirement. To the appeal for public contributions, Prince Kung gave Tls. 20,000. As the work became more extensive and demands for materials increased, the provinces became restive. Kung's memorial alluded to the rumor that the emperor was frolicking with the eunuchs while state papers were left unopened. During the audience that followed the presentation of the memorial Prince Ch'un gave proof of the emperor's private visits outside the palace by naming the times and the places, which made the emperor even angrier. See ibid., pp. 211-214.

36. Ibid., p. 226.

37. Ibid., pp. 226, 209-211.

38. Ibid., pp. 233-236. Work was resumed in the 1880s and 1890s.

39. Ibid., pp. 127, 129 ff, 145-47.

40. S. Wright, p. 405 ff. See *WCSL,* 1:28b, 3lb, chüan 4, 5, passim; also, *BPP,* China No. 1-c. 1422 (1876), China, No. 3-c. 1832 (1877).

41. Wade to Kung, Aug. 21, 1875, *BPP,* China No. 1-c. 1422 (1876), p. 61 China No. 3-c. 1832 (1877), p. 147; *WCSL,* 6:6-11.

42. Li Hung-chang, *Li Wen-chung-kung ch'üan-chi* (Shanghai, 1921), "I-shu han-Kao," chüan 3,4,5, passim.

43. Ibid., 3:32b, 33, 39; 4:8b; 5:18; *BPP,* China No. 1-c. 1422 (1876), pp. 89, 92; China No. 3-c 1832 (1877), pp. 105, 132-133.

44. Morse, *International Relations,* II, 303-304.

45. Ibid.

46. Li Hung-chang, 5:28b ff.

47. *WCSL,* 6:16b-17.

48. Li Hung-chang, 5:34b-35.

49. Wade to Derby, Oct. 27, 1875, *BPP* China No. 1-c 1422 (1876), p. 104.

50. Hart to Campbell, Aug. 24, 1876, quoted in S. Wright, p. 432 n. 4.

51. Li Hung-chang, 6:30.

52. Wade to Derby, July 14, 1877, *BPP,* China No. 3-c. 1832 (1877), p. 147.

53. S. Wright, p. 411.

54. Inspector General's Circular No. 709 (Apr. 30, 1896) in *Documents,* II, 55-56.

55. Born in Julich, Prussia, he joined the Chinese Customs in 1865. See footnote, *ibid.,* I, 402.

56. *Report of the Chinese Post Office (1921), pp. 34 ff., 105-106.*

57. Li Hung-chang, 8:18.

58. Summary of the commissioner's report, July 1879, in *Report of the Chinese Post Office (1921),* pp. 105-106.

59. *Report of the Chinese Post Office (1921),* pp. 34, 105.

60. *Decennial Reports, 1882-1891,* p. 73.

61. *Report of the Chinese Post Office (1921),* p. 34.

62. Ibid.

63. *Chung-kuo yu-p'iao mu-lu* (Taipei, 1956), p. 11.

64. *Report of the Chinese Post Office (1921),* appendix B. p. 106; Morse, *International Relations,* III, 63. Hart to Kopsch, Aug. 17, 1894. Hart Manuscripts.

65. Li Hung-chang, 8:18.

66. *Report of the Chinese Post Office (1921),* p. 34.

67. Inspector General's Circular No. 89 in *Documents,* I, 401-402.

68. *Report of the Chinese Post Office (1921),* p. 34.

69. Inspector General's Circular, No. 89 in *Documents,* I, 402.

70. Commissioner's report, July 1879, quoted in *Report of the Chinese Post Office (1921),* p. 105.

71. Ibid., p. 34.

72. *Documents,* I, 403.

73. *Decennial Reports, 1882-1891,* p. 74.

74. *Decennial Reports, 1892-1901,* I, 32.

## VI. *Attempts to Create a National Post Office*

1. *NCH* (July 4, 1884); cf. Hart to Charles Hannen, June 12, 1884, Hart Manuscripts.

2. *NCH* (Oct. 15, 1884).

3. Kuo Sung-t'ao (1818-1891), a scholar and statesman with enlightened views from Hsiang-yin, Hunan, was for some time acting governor of Kwangtung and judicial commissioner of Fukien before his appointment to England. See Hummel, I, 438-439.

4. *CKHP, wai-pien,* 18:9 a-b.

5. Ibid., 18:9b-12. In the early days one minister was appointed to several countries, and very often a joint-minister was appointed at the same time.

6. Ibid.

7. Ibid.

8. Ibid., 18:11.

9. Li Hung-chang, *Memorials* 37:30.

10. *CKHP, wai-pien,* 18:12-14.

11. Ibid., 18:13b-14.

12. Li Hung-chang, 37:30.

13. Morse, *Trade and Administration,* p. 411; Chang Liang-jen, *Chung-kuo yu-cheng* (Shanghai: Commercial Press, 1935-1936), I, 11. See "Canton," *Decennial Reports, 1882-1891,* p. 573; "Foochow," and "Kiungchow," *Decennial Reports, 1892-1901,* II, 111, 392. The bureau in Canton, established in 1882, had an agency in Hongkong, and Governor-General Tseng Kuo-ch'üan used a secret code worked out by Marquis Tseng Chi-tse for telegraphic communications with Tsungli Yamen. Tseng Kuo-ch'üan, *Tseng Chung-hsiang-kung shu-cha* (N. p. , 1903), 17:2.

14. *CKHP, wai-pien,* 18:12b-13.

15. *Decennial Reports, 1882-1891,* p. 573.

16. *Tung-hua hsü-lu,* Kuang-hsü 15 (1889), 97:2.

17. Ibid., Kuang-hsü 20 (1894), 123:5.

18. Ch'ü Hung-chi, Kuang-hsü 17. 9.13-16, 22-23 (Oct. 15-18, 24-25, 1891).

19. Ibid., Kuang-hsü 17.6.25 to 17.8.1, (July 30 to Sept. 3, 1891) and Kuang-hsü 17.9.26-29, 17. 10.4-6 (Oct. 28-30, Nov. 5-7, 1891), with one week's stay in Shanghai.

20. Hua Hsüeh-lan, *Hsin-ch'ou jih-chi,* ed. T'ao Meng-ho (Shanghai: Commercial Press, 1936). For Lu P'ei-fen see *Pei-chuan chi-pu* (Peiping, 1932), 20:13b-15.

21. Ch'ü Hung-chi, Kuang-hsü 17.5.4, 6, 14, 16, 21 (June 10, 12, 20, 22, 27, 1891), pp. 9-13.

22. Yen Hsiu, *Yin-hsiang-kuan shih-Ch'ien jih-chi* (Tientsin, 1935).

23. Hua Hsüeh-lan, pp. 63, 75, 102-171.

24. Ibid., pp. 184-185.

25. Ibid., pp. 180, 184, 185, 189, 195-197; See notes on pp. 200, 202.

26. Ibid., pp. 48, 71, 100, 201.

27. Ibid., p. 206-207.

28. *Tung-hua hsü-lu*, 104:14b-19b.

29. Ibid.

30. Ibid., 105:10-11.

31. Hummel, I, 526-528. H. B. Morse, then acting commissioner of
Customs at Tamsui, said that Liu's work was so beneficial on the whole that
"when he left in June 1891, the general feeling was in losing Commissioner
Liu, Formosa lost a part of itself." *Decennial Reports, 1882-1891*, pp. 449-450,
460.

32. Liu Ming-ch'üan's memorial, December 1889, in *Tung-hua hsü-lu*,
97:3. Ch'en Ming-chih was appointed as circuit intendent.

33. *Decennial Reports, 1882-1891*, pp. 488-489.

34. *Tung-hua hsü-lu*, Kuang-hsü 15 (1889), 97:3.

35. The Yamen's memorial of Kuang-hsü 22. 2.7 (Mar. 20, 1896) for the
establishment of the post office gives a brief account of various efforts made
toward the project before 1896. See *CTLT*, 463: 1b-2b; also, *Huang-ch'ao
Tao-Hsien-T'ung-Kuang tsou-i*, comp. Wang Yen-hsi and Wang Shu-min
(Shanghai, 1902), 14:7b.

36. Hummel, I, 331-333. Hsüeh had served on the staff of Tseng Kuo-fan
and Li Hung-chang, and in 1889 he was appointed minister to England, France,
Italy, and Belgium.

37. Kopsch became the first Postal Secretary in 1896. He joined the
Customs service in 1862 and served in Kiukiang, Chinkiang, Ningpo, Tainan,
Shanghai, Newchwang, and other places as commissioner before occupying
the post of Statistical Secretary in Shanghai (1891-1897). Note to the
Inspector General's Circular No. 709 in *Documents*, II, 56.

38. *CTS:YCP*, 1:7-10.

39. *CTLT*, 463:2.

40. Hsüeh Fu-ch'eng, "Ch'uang-k'ai Chung-kuo t'ieh-lu," *Yung-an ch'üan-
chi* (Shanghai, 1897), *wen-pien*, 2:11b, 15.

41. *CTS:YCP*, 1:12; the Yamen's memorial, Kuang-hsü 22. 2.7, *CTLT*,
463:1b-2.

42. *CTS:YCP*, 1:7-9.

43. *CTLT*, 463:2.

44. Ibid.

45. See Hart's discussion with Li Hung-chang in 1876 during the Margary
crisis and with the Yamen in 1890 and 1895. Inspector General's Circular, No.
709 (Apr. 30, 1896) in *Documents*, II, 55.

46. "Liu-tsei: Chang Hsien-chung, Li Tzu-ch'eng," *Ming-shih, lieh-chuan*, 197:3b.

47. Ibid., 2b-3.

48. Cheng Kuan-ying, *Sheng-shih wei-yen* (Canton, 1900), 7:72, and other contemporary essays.

49. Kopsch's immediate successor was Kleinwachter. The reports were made by Merrill in 1891 for the *Decennial Reports, 1882-1891*, p. 382.

50. Ibid.

51. S. Wright, pp. 629-630.

52. Ibid., chaps. 18, 19, passim; Morse, *International Relations*, vol. 2, chap. 18; Teng and Fairbank, *China's Response to the West*, pp. 123-124.

53. S. Wright, chaps. 18, 19, esp. p. 479 ff.

54. Hart to Kopsch, June 18, 1889, Hart Manuscripts.

55. *CTLT*, 463:2.

56. S. Wright, p. 630.

57. Ibid.

58. Ibid.

59. Ibid., pp. 630-631; there were thirteen articles of a general nature in the outline of the plan. See *CTS:YCP*, I, 15.

60. Hart to Kopsch, Dec. 15, 1892, Hart Manuscripts.

61. Hart to Kopsch, Mar. 24, 1893, ibid.

62. According to one report, the number of Westerners to be engaged was not to exceed 600. See *T'oung pao*, vol. 5, p. 63 (1894).

63. *HT:KH*, 6:13.

64. S. Wright, p. 631.

65. See memorials of Wen-hsiang, 1:2b and Li Hung-chang, *I-shu han-kou* 11:5.

66. S. Wright, pp. 599-601. Hart declined because it was difficult to find a successor. His home leave would have been effective in 1886, but in 1889 he was still in China. The Sikkim affair was the latest reason for postponing his leave.

67. Hart to Charles Hannen, June 17, 1887; Hart to Kopsch, June 18, 1889; Hart to C. Hennen, July 8, 1893, in Hart Manuscripts.

68. Julius M. Price of *The World*, "A Talk with Sir Robert Hart," *NCH* (Sept. 21, 1894).

69. Lo Yü-tung, "Kuang-hsü ch'ao pu-ch'iu ts'ai-cheng chih fang-ts'e," *Chung-kuo chin-tai ching-chi yen-ch'iu chi-k'an*, 1.2: 189-191 (May 1933); Hu Chün, *Chung-kuo ts'ai-cheng chih chiang-i* (Shanghai, 1920), chap. 9, p. 332 ff.

70. S. Wright, p. 632.

71. Wu Hsiang-hsiang, pp. 205-206, 233-236.

72. Lo Yü-tung, pp. 211-213.

73. Ibid., pp. 192-210.

74. Ibid., pp. 263-270.

75. Hart to Kopsch, Dec. 15, 1892, Hart Manuscripts.

76. S. Wright, p. 632.

77. Bishop, Morton, and Sayers, pp. 123-125; *NCH* (July 27, 1894; Aug. 17, 1894; Ichang, December 1894).

78. Hart to Kopsch, June 28, 1893, Hart Manuscripts.

79. The Yamen's memorial Mar. 20, 1896, *CTLT*, 463: 1b-2.

80. Hart to Kopsch, Aug. 21, 1893, Hart Manuscripts.

81. Ibid.

82. S. Wright, p. 631.

83. The note should be accompanied by an authentic French translation and "delivered officially to the Swiss foreign office by a Chinese diplomatic agent accredited to one or other of the European Powers . . . " Ibid.

84. Hart to Kopsch, Aug. 21, 1893, Hart Manuscripts.

85. S. Wright, p. 631.

86. Hart to Kopsch, Feb. 13, 1894, Hart Manuscripts.

87. Li Chieh-nung, *The Political History of China, 1840-1928,* tr. and ed. Teng Ssu-yu and Jeremy Ingalls (Princeton: Van Nostrand, 1956), p. 132 ff.

88. In August 1894 Hart thought that postal reform was going through but suddenly the Yamen dropped the plan. "Its opponents are therefore stronger than its supporters, but time will probably bring it out again . . . " Hart to Kopsch, Nov. 17 and Nov. 30, 1894; June 21, 1895. Hart Manuscripts

VII. *The Establishment of the Imperial Post Office*

1. Feng died in 1874, but this essay mentioned the China Merchants Steam Navigation Company and the Peking-Tientsin Railroad, established in 1876 and 1881 respectively. In any case, the latter was not even sanctioned before Feng's death. Moreover, the essay was not dated and not found in any of Feng's collected works. See "Ts'ai I-chan i," *WHTK*, 375:16; Feng Kuei-fen, pp. 62-66, 67-70 for essays on adoption of Western technology. See Teng and Fairbank, *China's Response to the West*, pp. 50-51.

2. *Huang-ch'ao ching-shih-wen, san pien,* ed. Ch'en Chung-i (Shanghai: Shang-hai shu-chü, 1901), chüan 55.

3. Cheng Kuan-ying, 7:63-73. See prefaces for Cheng's career. His essays "Yu-cheng," and "T'ai-hsi yu-cheng k'ao" were printed in several editions of *Ching-shih-wen.*

4. A secretary at one time for the Society for the Diffusion of Christian and General Knowledge among the Chinese, he was very eager to see China modernized. See William Soothill, *Timothy Richard.*

5. Cheng Kuan-ying, 7:63b-64.

6. Ibid., 7:63-73.

7. Essays of Lo Yü-lin and Wang I-san in *Huang-ch'ao ching-shih-wen, san-pien* chüan, 55; T'ang ch'en, *Wei-yen* (Warnings) 2:29-33: Teng and Fairbank, *China's Response to the West*, p. 112.

8. *Decennial Reports, 1882-1891*, p. 315; *Decennial Reports, 1892-1901*, II, 606; Yang Kung-hao, pp. 110-111.

9. Teng and Fairbank, *China's Response to the West*, chaps. 14-17, passim; Morse, *International Relations*, III, 55-56.

10. Hart to Kopsch, Aug. 29, 1895, quoted in ibid., p. 65 n. 12.

11. For Li and Weng, see Hummel. See also, Hsiao Kung-ch'üan, "Weng T'ung-ho and the Reform Movement of 1898," *Ch'ing-hua hsüeh-pao*, no. 1, pt. 2, pp. 119-121, 126 (April 1957).

12. Hart to Kopsch, Aug. 29, 1895, quoted in Morse, *International Relations*, III, 65 n. 12.

13. Hart to Kopsch, Sept. 9, 1895, ibid., n. 13.

14. Liu was called to take command of the troops at Shanhaikuan during the crisis of 1894.

15. Hart to Kopsch, Aug. 29, 1895, in Morse, *International Relations*, III, 65 n. 12.

16. Memorial of Kuang-hsü 21.11.12 (Dec. 27, 1895) in *Chang Wen-hsiang-kung ch'üan-chi, tsou-i*, 40:9b.

17. The other two projects were the training of a new army according to Western methods and the building of the Kiangsu-Chekiang railway. Ibid 40:8b-10b.

18. The Tsungli Yamen received a communication about the rescript on Jan. 17, 1896 (Kuang-hsü 21.12.3), *CTLT*, 463:1b.

19. *CTLT*, 463:1.

20. *CTLT*, 463:1b-2b.

21. *CTLT*, 463:8-11.

22. Ch'i Ju-shan, "Man-t'an Chung-kuo yu-shih," *Ch'i Ju-shan sui-pi* pp. 34-35.

23. Anonymous essays and reprints of newspaper articles and editorials written between 1895 and 1897 in *Huang-ch'ao ching-shih wen, t'ung-pien* ed. Jun Fu (Shanghai: Pao-shan ch'ai, 1901), chüan 94.

## VIII. *Retardation of Modernization and Its Consequences*

1. Inspector General's Circular No. 706 (Apr. 9, 1896) in *Documents*, II, 42.

2. Inspector General's Circular No. 709 (Apr. 30, 1896) in ibid., p. 55

3. Ibid., p. 56.

4. *CTS: YCP*, I, 67; Chang Liang-jen, "Chung-kuo li-tai yu-chih kai-yao," 32.1:15-16.

5. Hart to Edward B. Drew, Oct. 20, 1907, Hart Manuscripts.

6. In 1922 there were 158 foreign postal agencies in China, over 100 of which belonged to Japan. See Bishop, Morton, and Sayers, pp. 168-175.

7. Ibid., pp. 158-164.

8. Hart to a friend, Nov. 25, 1903, Hart Manuscripts.

9. Postal receipts exceeded expenditure for the first time in 1915. *Report of the Chinese Post Office (1921), Documents*, VII, 290.

10. Ho Tsung-yen, "Yu-cheng cheng-ts'e chih chien-t'ao," *T'ieh-lu ch'i-shih-wu, tien-hsin ch'i-shih-wu, yu-cheng liu-shih chou-nien chi-nien-k'an* (Taipei: Ministry of Communications, 1956), *yu-cheng*, p. 8.

11. Hart to Charles Hannen, Dec. 11, 1898; Hart to E.B. Drew, Oct. 20, 1907, Hart Manuscripts; *Documents*, II, 184. For a resume of Hart's work, see ibid., pp. 628-663.

## Bibliography

An Yung-chi. *Das Postwesen in China und Seine Entwicklung.* Leipzig: University of Leipzig, 1930.

Andersch, Max. *Die Deutsch Post in der Turkei, in China und in Marokko.* Berlin: R.V. Decker's Verlag, 1912.

*Annual Report of the Postmaster General of the United States.* Washington, D.C.: 1867, 1868, 1869, 1870, 1872.

Arlington, L.C. *Through the Dragon's Eyes: Fifty Years' Experiences in the Chinese Government Service.* London: Constable, 1931.

Bishop, George Thompson, C.S. Morton, and W. Sayers. *Hong Kong and The Treaty* Ports: *Postal History and Postal Markings.* 2nd ed.; revised and enlarged by Harold Edward Lobdell and Adrian E. Hopkins, London: The Postal History Society, 1949.

Bredon, Juliet. *Sir Robert Hart: The Romance of a Great Career.* London: Hutchinson, 1909.

*British Parliamentary Papers:*
*Third Report of the Select Committee on Postage.* Reports of Committees, 1837-1838. Vol. 20.
China No. 1-c. 1422. "Correspondence Respecting the Attack on the Indian Expedition to Western China and the Murder of Mr. Margary." 1876.
China No. 4-c. 1605. Continuation of c. 1422. 1876.
China No. 3-c. 1832. Continuation of c. 1605. 1877.
China No. 2-c. 2716. "Correspondence Respecting the Agreement Between the Ministers Plenipotentiary of the Governments of Great Britain and China Signed at Chefoo on September 13, 1876." 1880.
China No. 3-c. 3395. Continuation of c. 2716. 1882.

Broomhall, Marshall, ed. *The Chinese Empire: A General and Missionary Survey.* London: China Inland Mission, 1908.

Brunnert, H.S. and V.V. Hagelstrom. *Present Day Political Organization of China,* tr. A. Beltchenko and E. E. Moran. Rev. ed.; Shanghai: Kelly & Walsh, 1912.

Cahen, Gaston. *Histoire des Relations de la Russie avec la Chine sous Pierre le Grand, 1689-1730.* Paris: Lib. Felix Alcan, 1912.

Chang Chih-tung 張之洞. *Chang Wen-hsiang-kung ch'üan-chi* 張文襄公全集 (Collected works of Chang Chih-tung). 229 chüan; Peiping: Wen-hua chai, 1928.

Chang Hao. "The Anti-Foreignist Role of Wo-jen (1804-1871)," *Papers on China,* 14:1-29 (1960). Harvard University, East Asian Research Center.

Chang Liang-jen 張樑任. "Chung-kuo li-tai yu-chih kai-yao" 中國歷代郵制概要(A historical survey of the postal system in China), *Tung-fang tsa-chih* 東方雜誌 (The eastern magazine), 32.1 (*yen* 研):7-16 (January 1935).

———*Chung-kuo yu-cheng* 中國郵政 (Postal administration in China). 3 vols.; Shanghai: Commercial Press, 1935-1936.

Chao Fung-t'ien 趙豐田. *Wan Ch'ing wu-shih-nien ching-chi ssu-hsiang shih* 晚清五十年經濟思想史 (Economic thought in the last fifty years of the Ch'ing dynasty). Yenching Hsüeh-pao Special, no. 18. Peiping: Harvard-Yenching Institute, 1939.

Ch'en Ch'i-t'ien 陳其田. *Shan-hsi p'iao-chuang k'ao-lüeh* 山西票莊考畧(A study of Shansi banks). Shanghai: Commercial Press, 1937.

Ch'en Fu-kuang 陳復光. *Yu-Ch'ing i-tai chih Chung-O kuan-hsi* 有清一代之中俄關係 (Sino-Russian relations during the Ch'ing dynasty). Yunnan University Law and Arts School series, B1. Yunnan, 1947.

Ch'en Tse-chuan (Ch'en Chih-ch'uan) 陳志川. *Chung-kuo tsao-ch'i yu-cheng ti shih-liao k'ao-shih* 中國早期郵政的史料考實 (A study of China's early postal history). Taipei: Cathay Philatelic Library, 1965.

Cheng Kuan-ying 鄭觀應. *Sheng-shih wei-yen* 盛世危言 (Warnings to a seemingly prosperous age). 8 chüan; enlarged ed.; Canton, 1900.

Ch'i Ju-shan 齊如山. "Man-t'an Chung-kuo yu-shih" 漫談中國郵史 (A cursory discourse on Chinese postal history), *Ch'i Ju-shan sui-pi.*

———"Piao-chü-tzu shih-hua" 鏢局子史話 (A historical sketch of the escort agencies), *Ch'i Ju-shan sui-pi.*

*Ch'i Ju-shan sui-pi* 齊如山隨筆 (Notes and reminiscences of Ch'i Ju-shan). Taipei: Chung-yang wen-wu kung-ying she, 1953.

*Chiao-t'ung shih: Yu-cheng pien* 交通史：郵政編(A history of communications: Postal service), comp. Committee for the Compilation of the History of Communications; chief editor, Chang Hsin-cheng 張心徵. 4 vols.; Nanking: Ministries of Communications and Railways, 1930.

Chin Liang 金梁 et al. *Chin-shih jen-wu chih* 近世人物志 (Notes on modern personages). N.p., 1934.

*Chine et Ceylan* (1898-1900).

"Ching-chi pien, yung-cheng tien, i-ti k'ao" 經濟編．戎政典,驛遞孝 (Economic affairs, military administration, and postal service), chüan 258-262 in *Ku-chin t'u-shu chi-ch'eng* 古今圖書集成 (Grand encyclopaedia or compilation of books and illustrations of ancient and modern times), ed. Chiang T'ing-hsi 蔣廷錫 et al. 10,000 chüan; Shanghai, 1884.

*Ch'ing-ch'ao hsü wen-hsien t'ung-k'ao* 清朝續文獻通攷 (Encyclo-paedia of the historical records of the Ch'ing dynasty), comp. Liu Chin-ts'ao 劉錦藻. 400 chüan; Shanghai: Commercial Press, 1936.

*Ch'ing-chi wai-chiao shih-liao* 清季外交史料 (Historical materials concerning foreign relations in the late Ch'ing period), comp. Wang Yen-wei 王彥威. 112 ts'e; Peiping, 1932-1935.

*Ch'ing-hua hsüeh-pao* 清華學報(Ch'ing-hua journal of Chinese studies). New Series, Taipei, 1957-.

*Ch'ing-shih kao* 清史稿 (A draft history of the Ch'ing dynasty), comp. Chao Erh-hsün 趙爾巽 et al., chüan 152. 536 chüan; Peking, 1927.

"Ch'ing-tai chou-hsien ku-shih" 清代卅縣故事 (Historical materials on local government in the Ch'ing dynasty), ed. Ts'ai Shen-chih 蔡申之, *Chung-ho yüeh-k'an* 中和月刊, 2.9:49-67 (September 1941); 2:10:72-95 (October 1941); 2.11:89-101 (November 1941), 2.12:100-108 (December 1941).

*Ch'ou-pan I-wu shih-mo* 籌辦夷務始末(The complete account of our management of barbarian affairs). Photolithograph of the original compilation; Peking: The Palace Museum, 1929-1931. Hsien-feng period (1851-1861), 80 chüan; T'ung-chih period (1862-1874), 100 chüan.

Chu Chang-hsing (Hsü Ch'ang-ch'eng 徐昌成 ). "Ku-tai yu-cheng shu-lüeh" 古代郵政述畧 (A survey of postal administration in ancient times), *Yu-i yüeh-k'an* 郵藝月刊 (Postal arts monthly), 1.1:5-6 (January 1947).

Ch'ü Hung-chi 瞿鴻襪. *Shih-Yü shih-Min jih-chi* 使豫使閩日記 (Diaries of the missions to Honan and Fukien), ed. Ch'ü Hsüan-ying 瞿宣穎. Peiping, 1933.

Ch'ü T'ung-tsu. *Local Government in China Under the Ch'ing*. Cambridge, Mass.: Harvard University Press, 1962.

*Chung-ho yüeh-k'an* 中和月刊 (Chung-ho monthly). Peiping, 1940-.

*Chung-hua yu-cheng yü-t'u* 中華郵政輿圖 (A postal atlas of China). Peking: Ministry of Communications, Directorate General of Posts. 1919.

*Chung-kuo yu-p'iao mu-lu* 中國郵票目錄 (A catalogue of Chinese stamps). Taipei: Ministry of Communications, Director General of Posts, 1956.

*Chung-O yüeh-chang hui-yao* 中俄約章會要 (Collection of treaties, conventions, etc. between China and Russia), comp. the Tsungli Yamen. Peking, 1882.


*Decennial Reports, 1882-1891*. First issue; Shanghai: Inspectorate General of Customs, 1893.

*Decennial Reports, 1892-1901*. Second issue; 2 vols.; vol. 1, Shanghai: Inspectorate General of Customs, 1904; vol. 2, Shanghai; Inspectorate General of Customs, 1906.

*Documents Illustrative of the Origin, Development, and Activities of the Chinese Customs service*. 7 vols.; Shanghai: Inspectorate General of Customs, 1937.


Fairbank, John K. *Trade and Diplomacy on the China Coast: The Opening of the Treaty Ports, 1842-1854*. 2 vols.; Cambridge, Mass.: Harvard University Press, 1953.

———and Teng Ssu-yü. "On the Transmission of Ch'ing Documents," *Harvard Journal of Asiatic Studies*. 4.1:12-46 (May 1939).

———"On the Types and Uses of Ch'ing Documents," *Harvard Journal of Asiatic Studies.* 5.1:1-71 (January 1940).

Feng Kuei-fen 馮桂芬. *Chiao-pin-lu k'ang-i* 校邠廬抗議 (Personal protests from the study of Chiao-pin). 2 ts'e; N.p., 1884.

*First Report of the Postmaster General on the Post Office.* London, 1855.

*Fourteenth Report of the Postmaster on the Post Office.* London, 1868.

Harlow, Alvin F. *Old Post Bags.* New York and London: Appleton, 1928.

Hart, Sir Robert. Manuscripts, letters and papers in Houghton Library, Harvard University.

———Correspondence with James D. Campbell. Letter Series A (semi-official), Letter Series Z (1868-1906) at the School of Oriental and African Studies, University of London.

———Typescript copies of his letters to Sir Francis Arthur Aglen, Nov. 26, 1888, to Nov. 14, 1911, at the School of Oriental and African Studies, University of London.

———*These From the Land of Sinim: Essays on the Chinese Question.* London: Chapman & Hall, 1901.

*Harvard Journal of Asiatic Studies.* Cambridge, Mass., 1936-.

Hemmeon, J.C. *The History of the British Post Office.* Harvard Economic Studies, no. 7. Cambridge, Mass.: Harvard University Press, 1912.

Hertslet, Godfrey E.P., ed. *China Treaties; Treaties etc. Between Great Britain and China and Between China and Foreign Powers.* Third ed., 2 vols.; London: Her Majesty's Stationery Office, 1908.

Hill, George Birbeck and Sir Rowland Hill. *Life of Sir Rowland Hill, K.C.B. and a History of the Penny Postage.* 2 vols.; London, 1880.

Hill, Sir Rowland. *Post Office Reform, Its Importance and Practicability.* London, 1837.

*Honan fu-i ch'üan-shu* 河南賦役全書 (A complete record of taxation and labor service in Honan, from Tao-kuang 16 [1836] to Kuang-hsü 7 [1881]). Rev. ed.; 122 ts'e; Honan, 1893.

*Honan ts'ai-cheng shuo-ming-shu* 河南財政說明書 (An explanatory statement and report of financial administration in Honan). Honan, 1909.

Hsiao Kung-ch'üan 蕭公權. "Weng T'ung-ho and the Reform Movement of 1898," Ch'ing-hua hsüeh-pao 清華學報, no. 1, pt. 2, pp. 111-245 (April 1957).

Hsieh Pao-chao. *The Government of China, 1644-1911*. Baltimore: Johns Hopkins University Press, 1925.

Hsieh Pin 謝彬. *Chung-kuo yu-tien hang-k'ung shih* 中國郵電航空史 ( A history of the post, the telegraph, and aviation in China). Shanghai: Chung-hua Press, 1946.

Hsü T'ung-hsin 許同莘 . *Chang Wen-hsiang-kung nien-p'u* 張文襄公年譜 (A chronological biography of Chang Chih-tung). Shanghai: Commercial Press, 1946.

Hsüeh Fu-ch'eng 薛福成. *Yung-an ch'üan-chi* 庸盦全集 (Collected works of Hsüeh-Fu-ch'eng). 12 ts'e; Shanghai, 1897.

Hua Hsüeh-lan 華學瀾. *Hsin-ch'ou jih-chi* 辛丑日記 (Diary of the year 1901), ed. T'ao Meng-ho 陶孟和. Shanghai: Commercial Press, 1936.

*Huang-ch'ao chang-ku hui-pien* 皇朝掌故彙編 (Collected historical records of the imperial dynasty), ed. Sung Ch'eng-chih 宋澄之 et al. 100 chüan; Shanghai, 1902.

*Huang-ch'ao cheng-tien lei-ts'uan* 皇朝政典類纂 (A classified compendium of administrative statutes of the imperial dynasty), ed. Hsi Yü-fu 席裕福. 500 chüan; Shanghai: T'u-shu chi-ch'eng chü, 1903.

*Huang-ch'ao ching-shih-wen hsin pien* 皇朝經世文新編 (A new compilation of essays on government and economics of the imperial dynasty), comp. Mai Chung-hua 麥仲華. Reprint; 21 chüan; Shanghai: Shang-hai i-shu-chü, 1901.

*Huang-ch'ao ching-shih-wen hsü-pien* 皇朝經世文續編 (Supplement to essays on government and economics of the imperial dynasty), comp. Ko Shih-chün 葛士濬. 120 chüan; Shanghai, 1896.

*Huang-ch'ao ching-shih-wen san-pien* 皇朝經世文三編 (Third compilation of essays on government and economics of the imperial dynasty), ed. Ch'en Chung-i 陳忠倚 , 80 chüan; Shanghai: Shang-hai shu-chü, 1901.

*Huang-ch'ao ching-shih-wen t'ung-pien* 皇朝經世文統編 (General collection of essays on government and economics of the imperial dynasty), ed. Jun Fu 潤甫. 107 chüan; Shanghai: Pao-shan chai, 1901.

*Huang-ch'ao Tao-Hsien-T'ung-Kuang tsou-i* 皇朝道咸同光奏議 (Memorials and public papers presented in the Tao-Kuang, Hsien-feng, T'ung-chih, and Kuang-hsü periods of the imperial dynasty), ed. Wang Yen-hsi 王延熙, and Wang Shu-min 王樹敏. 64 chüan; Shanghai, 1902.

Huang Liu-hung 黃六鴻. *Fu-hui ch'üan-shu* 福惠全書 (Advice for a beneficial administration). 32 chüan; Peking, 1893.

Hummel, Arthur W., ed. *Eminent Chinese in the Ch'ing Period, 1644-1912.* 2 vols.; Washington, D.C.: Library of Congress, 1943-1944.

*Hu-nan chiang-yü i-ch'uan tsung-ts'uan* 湖南疆域驛傳總纂 (Complete guide to the official postal system in Hunan and between Hunan and neighboring provinces), comp. under the auspices of the Board of War. Rev. ed.; 10 chüan; maps, 1 chüan; 1888.

"I-chan lu-ch'eng" 驛站路程 (Post routes of the official mounted courier service), in *Hsiao-fang-hu-chai yü-ti ts'ung-ch'ao* 小方壺齋輿地叢鈔 (Collection of geographical works from the Hsiao-fang-hu studio), ed. Wang Hsi-ch'i 王錫祺, 1.3:184-187. Preface dated 1877; Shanghai, n.d.; also in *Tsai pu-pien* 再補篇 (Second supplement), vol. 1, no. 71.

"Inland Communications in China," ed. T.W. Kingsmill, *Journal of the China Branch of the Royal Asiatic Society,* New Series, 28:1-128 (1893-1894).

King, Paul. *In the Chinese Customs Service.* Rev. ed.; London: Heath Cranton, 1930.

Ko Kung-chen 戈公振. *Chung-kuo pao-hsüeh-shih* 中國報學史 (A history of journalism in China). Shanghai: Commercial Press, 1927.

Legge, James. "The Colony of Hongkong: A Lecture on Reminiscences of a Long Residence in the East," *The China Review,* 1.3:163-176 (November-December 1872).

Li Chien-nung 李劍農. *Chung-kuo chin-pai-nien cheng-chih shih* 中國 近百年政治史 (The political history of China in the last one hundred years). 2 vols.; Shanghai: Commercial Press, 1947.

———*The Political History of China, 1840-1928,* tr. and ed. Ssu-yü Teng and Jeremy Ingalls. Princeton: Van Nostrand, 1956.

Li Hsi-sheng 李希聖 , ed. *Kuang-hsü k'uai-chi lu* 光緒會計錄 (Financial records of the Kuang-hsü reign—based on reports of the year 1893), 3 chüan: Shanghai, 1896.

Li Hung-chang 李鴻章 . *Li Wen-chung-kung ch'üan-chi* 李文忠公全集 (The collected papers of Li Hung-chang). 166 chüan; Shanghai, 1921. See especially *I-shu han-kao* 譯署函稿 (Correspondence with the Tsungli Yamen), chüan 3-6, 8.

Li Kuang-t'ao 李光濤. "Ming-chi i-tsu yü liu-tsei" 明李驛卒與流賊 (Couriers and freebooters under the late Ming), *Ch'ing-chu Li Chi hsien-sheng ch'i-shih-sui lun-wen chi* 慶祝李濟先生七十歲論文集 (Symposium in honor of Dr. Li Chi on his seventieth birthday), *Ch'ing-hua hsüeh-pao*, pt. 2, pp. 807-830 (January 1967).

Li Sung-p'ing 李頌平. *K'e-yu wai-shih* 客郵外史 (A history of foreign postal establishments in China), *Chung-kuo yu-hsüeh ts'ung-shu* 中國郵學叢書 (Series on Chinese philatelic studies). Hong Kong: Pao-an yu-piao-she, 1966.

Lin Tse-hsü 林則徐 "Lin tse-hsü jih-chi" 林則徐日記 (Lin Tse-hsü's diary—excerpts from the years 1839, 1840, and 1841), *Ya-p'ien chan-cheng* 鴉片戰爭 (The Opium War). 6 vols.; Shanghai: Shen-chou kuo-kuang-she, 1954. Sources on Modern Chinese History Series, No. 1 ed. Ch'i Ssu-ho 齊思和, Lin Shu-hui 林樹惠, and Shou Chi-yü 壽紀瑜.

*Liu-pu ch'u-fen tse-li, li* 六部處分則例, 吏 (Rules and precedents of administrative decisions of the Six Boards for the punishment of government officials and employees, section on the Board for Civil Appointment). 50 chüan; 1887.

Liu Yüeh-yün 劉嶽雲, comp. *Kuang-hsü k'uai-chi piao* 光緒會計表 (Financial reports, Kuang-hsü 11 to 20 [1885-1894]). 4 ts'e; Chiao-yü shih-chieh-she, 1901.

Lo Yü-tung 羅玉東. "Kuang-hsü ch'ao pu-chiu ts'ai-cheng chih fang-ts'e" 光緒朝補救財政之方策 (Governmental policies for meeting the financial crises during the Kuang-hsü period), *Chung-kuo chin-tai ching-chi yen-chiu chi-k'an* 中國近代經濟研究集刊(Journal of economic studies of modern China), 1.2:189-191 (May 1933).

Lobdell, H.E. and A.E. Hopkins. See G.T. Bishop, C.S. Morton, and W. Sayers.

Lou Tsu-i 樓祖詒. *Chung-kuo yu-i fa-ta shih* 中國郵驛發達史 (A history of the development of postal service in China). Kunming: Chung-hua Press, 1940.

———*Chung-kuo yu-i shih-liao* 中國郵驛史料(Historical materials on Chinese postal history). Peking: Jen-min yu-tien ch'u-pan she, 1958.

Martin, W.A.P. *A Cycle of Cathay.* 2nd ed.; New York: Fleming H. Ravell, 1897.

Meng Ssu-ming. "The Organization and Functions of the Tsungli Yamen." Ph.D. thesis, Harvard University, 1949.

Momose Hiromu 百瀨弘. "Shimmatsu no Keisei bumhen ni tsuite" 清末の經世文編について(On the *Ching-shih wen-pien* of the late Ch'ing), *Ikeuchi hakushi kanreiki kinen Tōyōshi ronsō* 池田博士還曆紀念東洋史論叢(Essays on Oriental history collected in commemoration of the sixtieth birthday of Dr. Ikeuchi), pp. 877-892. Tokyo: Zanho Kan Kokai, 1940.

———"Feng Kuei-fen chi ch'i chu-shu" 馮桂芬及其著述 (Feng Kuei-fen and his writings), *Chung-ho yüeh-k'an,* 3.3:35-66 (March 1942).

Morse, H.B. *International Relations of the Chinese Empire.* 3 vols.; New York and London: Longmans & Green, 1910-1918.

———*Trade and Administration of the Chinese Empire.* Shanghai: Kelly & Walsh, 1908.

*Mu-tsung I-huang-ti sheng-hsün* 穆宗毅皇帝聖訓 (Sacred instructions of Emperor Mu-tsung), in *Shih-ch'ao sheng-hsün* 十朝聖訓 (Sacred instructions of ten reigns), comp. 1686-1879. Peking, n.d.

*North China Herald.* Shanghai, 1850-.

*O-sheng chou-hsien I-ch'uan ch'üan-t'u* 鄂省州縣驛傳全圖(Maps of postal communication in Hupeh). 2 ts'e; N.p., n.d.

Pai Shou-i 白壽彝 . *Chung-kuo chiao-t'ung-shih* 中國交通史 (A history of communications in China). Series on Chinese cultural history. Shanghai: Commercial Press, 1937.

*Ping-pu chung-shu cheng-k'ao* 兵部中樞政考(References for the central administration of the Board of War). Rev. ed.; 40 chüan; Peking, 1825; supplements, 4 chüan; 1832.

Piry, Théophile. "Report on the Working of the Post Office, 1904," *Decennial Reports, 1892-1901,* p. 314.

Polo, Marco. *The Travels of Marco Polo the Venetian,* tr. William Marsden. Everyman ed.; London: J. M. Dent, 1923.

*Post Circular, The Advocate for a Cheap, Swift and Sure Post* (July 5, 1838). Reprint; *London and Westminister Review* (April 1938).

*Report on the Chinese Post Office for the Tenth Year of Chung-hua Min-kuo (1921), with which is Incorporated an Historical Survey of the Quarter-century, 1896-1921.* Shanghai: Ministry of Communication, Directorate General of Posts, 1922.

Robinson, Howard. *The British Post Office: A History.* Princeton: Princeton University Press, 1948.

———*Britain's Post Office.* Oxford: Oxford University Press, 1953.

Schafer, Edward. *The Golden Peaches of Samarkand: A Story of T'ang Exotic.* Berkeley: University of California Press, 1963.

Shen Kuei-fen 沈桂芬."Shen Wen-ting Yüeh-yao jih-chi" 沈文定粵 輶日記(Shen Kuei-fen's diary of the mission to Canton), *Chung-ho yüeh-k'an,* 1.4:95-103 (April 1940), 1.5:79-84 (May 1940), 1.6:92-98 (June 1940).

Shen Ping-yüan 沈炳垣. *Shen Wen-chieh-kung hsing-yao jih-chi* 沈文 節公星輶日記 (The diary of a mission [to Szechwan], 1853). Preface, 1879; Szechwan, 1885.

*Shuo-wen chieh-tzu ku-lin* 說文解字詁林(Collected commentaries to Hsü Shen's 許慎 *Shuo-wen chieh-tzu* [Dictionary of Etymology]), comp. Ting Fu-pao 丁福保 . Reprint; Shanghai, 1937.

Smith, A. D. *The Development of Rates of Postage.* Studies in Economic and Political Science Series, no. 50. London: Allen and Unwin, 1917.

*Tables of Post Offices in the United States.* New York: U.S. Post Office, 1819.

*Ta-Ch'ing chin-shen ch'üan-shu* 大清搢紳全書 (A complete directory of Ch'ing dynasty officials). Quarterly publication, about 6 chüan per issue; Peking.

*Ta-Ch'ing hui-tien* 大清會典 (Collected statutes of the Ch'ing dynasty). Chia-ch'ing ed. (1818), 80 chüan; Kuang-hsü ed. (1899), 100 chüan.

*Ta-Ch'ing hui-tien shih-li* 大清會典事例 (Supplementary cases to the collected statutes of the Ch'ing dynasty), Chia-ch'ing ed. (1818), 920 chüan; Kuang-hsü ed. (1899), 1220 chüan.

*Ta-Ch'ing lü-li* 大清律例 (Laws and precedents of the Ch'ing dynasty). Rev. ed.; 47 chüan; supplement, 5 chüan; 1870.

*T'ang liu-tien* 唐六典 (Six statutes of the T'ang dynasty), ed. Li Lin-fu 李林甫 et al. under the auspices of Emperor Hsün-tsung. 30 chüan; Shanghai, 1895.

T'ao Pao-lien 陶保廉. *Hsin-mou shih-hsing chi* 辛卯侍行記 (A record of the journey attending my father in 1891). 6 chüan; Peking, 1896.

T'eng Ssu-yü and John K. Fairbank. *China's Response to the West: A Documentary Survey, 1839-1923.* 2 vols.; Cambridge, Mass.: Harvard University Press, 1954.

*T'ieh-lu ch'i-shih-wu, tien-hsin ch'i-shih-wu, yu-cheng liu-shih chou-nien chi-nien k'an* 鐵路七十五，電信七十五，郵政六十週年紀念刊 (A com.-memorative volume on the seventy-fifth anniversary of the railways and the telegraph and the sixtieth anniversary of postal administration). Taipei: Ministry of Communications, 1956.

Tōa Dobunkai 東亞同文會 , comp. *Shina keizai zensho* 支那經濟全書 (China economic series), vol. 6. Vols. 1-4, Osaka, 1907; vols. 5-12, Tokyo, 1908.

*Treaties, Conventions, etc. Between China and Foreign States.* 2 vols.; Shanghai: Inspectorate General of Customs, 1917.

140

Tseng Kuo-fan 曾國藩. *Tseng Wen-cheng-kung chia-shu* 曾文正公家書 (Tseng Kuo-fan's family letters). Shanghai: Commercial Press, 1905.

*Tso-chuan* 左傳 (Tso Ch'iu-ming's commentaries to the *Spring and Autumn Annals)*. N.p., n.d.

T'u Ssu-ts'ung 唐思聰. *Chung-hua tsui-hsin hsing-shih t'u* 中華最新形勢圖 (The latest atlas of China). Shanghai: Shih-chieh yü-ti hsüeh-she, 1933.

*Tung-hua hsü-lu* 東華續錄 (The Tung-hua records), ed. Chu Shou-p'eng 朱壽朋, Kuang-hsü period (1875-1908), Shanghai, 1909.

Wang Ch'ing-yün 王慶雲. *Shih-ch'ü yü-chi* 石渠餘紀 or *Hsi-ch'ao chi-cheng* 熙朝紀政 (The administration in the Ch'ing dynasty). 6 chüan; Hunan, 1890.

Wang K'ai-chieh 王開節. *Wo-kuo yu-cheng fa-chan chien-shih* 我國郵政發展簡史 (A brief history of the development of our postal system). Taipei: Chung-kuo chiao-tung chien-she hsüeh-hui, 1954.

———*Chung-kuo chin-pai-nien chiao-tung-shih* 中國近百年交通史 (A history of communications in China in the last one hundred years). Taipei: Chung-kuo chiao-tung chien-she hsüeh-hui, 1960.

Wei Chü-hsien 衛聚賢. *Shan-si p'iao-hao shih* 山西票號史 (A history of Shansi banks). Chungking: Department of Economic Studies of the Central Bank of China, 1944.

Wen-hsiang 文祥. *Wen Wen-chung-kung shih-lüeh* 文文忠公事略 (The papers of Wen-hsiang). 4 ts'e; N.p., 1822.

Weng T'ung-ho 翁同龢. *Weng Wen-kung-kung jih-chi* 翁文恭公日記 (Weng T'ung-ho's diary). 40 ts'e; Shanghai: Commercial Press, 1925.

Wright, Mary. *The Last Stand of Chinese Conservatism: The T'ung-chih Restoration, 1862-1874.* Stanford: Stanford University Press, 1957.

Wright, Stanley F. *Hart and the Chinese Customs.* Belfast: William Mullen & Son for Queen's University, Belfast, 1950.

Wu Hsiang-hsiang 吳相湘. *Wan-Ch'ing kung-t'ing shih-chi* 晚清宮廷實紀 (True accounts of the imperial court in the late Ch'ing period taken from palace archives). First series. Taipei: Cheng-chung shu-chü, 1952.

*Wu-hsü pien-fa* 戊戌變法 (Reform of 1898), comp. Chien Po-tsan 翦伯贊 et al. *Chung-kuo chin-tai-shih ts'ung-k'an* 中國近代史叢刊 (Sources on modern Chinese history series), ed. Chinese Historical Association. 4 vols.; Shanghai: Shen-chou kuo-kuang-she, 1953.

Yao Kung-hao 姚公鶴. "Shang-hai pao-chih hsiao-chih" 上海報紙小史 (A brief history of Shanghai newspapers), *Shang-hai hsien-hua* 上海閒話 (Reminiscences of Shanghai). Shanghai: Commercial Press, 1926.

Yen Hsiu 嚴修. *Yin-hsiang-kuan shih-Ch'ien jih-chi* 蟫香館使黔日記 (Diary of the mission to Kweichow). 9 ts'e; Tientsin, 1935.

# GLOSSARY

Aksu 阿克蘇

Altai 阿爾泰

An-ch'a-shih 按察使

An-ting 安定

Anking 安慶

Barkol 巴里坤

chan 站

Chang-chia 張家

chang-ching 章京

Changshuchen 樟樹鎮

Changteh 常德

Ch'ang-yih (Ch'ang-yeh) 長掖

*Chao-wen hsin-pao* 昭文新報

Chaotung 昭東

Ch'ao-ch'ü 朝渠

Ch'ao-yang 朝陽

Ch'e-chia ch'ing-li ssu 車駕清吏司

ch'en 辰

Ch'en-chia 陳家

Ch'en Ming-chih 陳鳴志

Cheng 鄭

Cheng-yang 正陽

Chengtu 成都

Ch'eng-lin 成林

Chian 吉安

"Chi-sheng yü Ch'ing Tao-Hsien-T'ung-Kuang chih-chiao" 極盛於清道咸同光之交

Chi T'ung 吉通

Ch'i-ho 齊河

Ch'i-men 祈門

Chieh-pao ch'u 捷報處

"Chien Shu chien hsi Cheng" 蹇叔諫襲鄭

chien-tu 監督

chih 置

chih-li-chou 直隸州

chih-li-t'ing 直隸廳

"Chih yang-ch'i i" 製洋器議

Ch'ih-shih 赤石

Chin 晉

*Ching-pao* 京報

Ch'ing chiang p'u 清江浦

Ch'ing-shan 青山

Ch'ing-shui 青水

*Ch'ing-tai t'ung-shih* 清代通史

Ch'ing-yün 青雲

Chingtechen 景德鎮

Chinkiang 鎮江

Chou 周

chou 州

Chou Heng-ch'i 周恒祺

Chu-sha 竹沙

Ch'u 楚

ch'uan 傳

"Chuang-k'ai Chung-kuo t'ieh-lu" 創開中國鐵路

*Chung-kuo ts'ai-cheng chih chiang-i* 中國財政之講義

*Chung-wai hsin-pao* 中外新報

Chü 遽

chün-t'ai 單台

Fancheng 樊城
fu 府
fu-chu-k'ao 副主考
Fu-lin 福林
Fuchow 涪州

Hai-kuan Po-ssu-ta shu-hsin-kuan
海關撥駟達書信館
Ho-ch'ang 合昌
Ho-chien fu 河間府
Ho Ju-chang (Tzu-o)
何如璋 (子峩)
Ho Tsung-yen 何縱炎
Hokowchen 河口鎮
Hsi-feng-k'ou 喜峯口
Hsi-liu 西流
Hsiang-yin 下馬
Hsiang-yin 湘陰
Hsiao I-shan 蕭一山
hsien 縣
Hsin-an 新安
hsin-chü 信局
hsün-ch'eng-ma 巡城馬
Hsün-huan jih-pao 循環日報
Hu ch'ü 胡渠
Hu Chün 胡鈞
Hu Lin-i 胡林翼
hu-p'ai 護牌
Hu-wan-ch'ang 胡萬昌
Hu Yü-fen 胡熵芬
Hua-tzu jih-pao 華字日報
Hua-yang shu-hsin kuan
華洋書信館
Huang-hua 黃花

Huang-hua i 皇華驛
Huang Hui-ho 黃惠和
Huang-ts'un 黃邨
Hui-pao 滙報
Hui-t'ung-kuan 會同館
Hukow 湖口
Hung-men 紅門
huo-p'ai 火牌
huo-p'iao 火票

i 驛
I-chan 驛站
Ichang 宜昌
I-cheng wang 議政王
i-ch'eng 驛丞
i-ch'uan 驛傳
I-hsin 奕訢
I-huan 奕環
I-k'uang 奕匡
I-ning-chou 義寧州
I-ti 驛遞
i-tsu 驛卒
Ili 伊犁
Iyang 益陽 (in Hunan)
Iyang 弋陽 (in Kiangsi)

Jaochow 饒州
Jen-chia 任家
Jih 馹
Jih-sheng-ch'ang 日昇昌

Kanchow 贛州
K'an-ho 勘合
Karashar 喀喇沙
Kashgar 喀什噶爾

Khotan 和闐
Kiangkowchen 江口鎮
Kiukiang 九江
"Ko-hang shih-chien" 各行事件
Kobdo (or *Jirgalanta*) 科布多
Ku-pei k'ou 古北口
Ku-ts'ung 谷㵗
K'u-lun pan-shih ta-ch'en

　庫倫辦事大臣
Kuche 庫車
K'uei-pin 奎斌
Kung-men-ch'ao 公門抄
"Kung-men yao-lüeh"

　公門要略
Kung-sun Ch'ou 公孫丑
kung-tu 公牘
Kuo Sung-t'ao 郭嵩燾
Kweichow 夔州
Kweiki 貴溪
Kweiyang 貴陽

Lai-chou fu 萊州府
Lanchow 蘭州
Laohokow 老河口
Laoyatan 老鴉灘
li 里
Li-fan yüan 理藩院
Li-ho 裏河
Li Hung-tsao 李鴻藻
Li K'uei 李圭
Li-pu 吏部．禮部
*Liang-hsiang pao* 良鄉報
Liang-kiang 兩江
lieh-chuan 列傳
Lin-p'in 臨品

Liu K'un-i 劉坤一
Liu Ming-ch'uan 劉銘傳
"Liu-ts'e: Chang Hsien-chung
　Li Tzu-ch'eng"

　流賊張獻忠李自成
Liu yang 瀏陽
Loping 樂平
Lo Yü-lin 羅毓林
Loyang 略陽
Lü P'ei-fen 呂珮芬
lun-ch'uan hsin-chü 輪船信局

Ma-an 馬鞍
Mao-t'ang 茅塘
min hsin-chü 民信局
Ming-huang 明皇
*Ming-shih* 明史
Mu-tsung 穆宗

Nanchang 南昌
Nanyang 南陽
nei-ko hsüeh-shih 內閣學士
Nieh Ch'i-kuei 聶緝槼
Nien-fei 捻匪
nien-p'u 年譜
Ning hsiang 寧鄉
Niu-chiao 牛角

Pai-mao 白茅
Pai-ts'un 白邨
p'ai-tan 排單
*Pei-chuan chi-pu* 碑傳集補
piao-chü 鏢局
p'iao-chuang 票莊
Ping-pu huo-p'iao 兵部火票

Ping-pu ssu-yüan 兵部司員

P'ing-kiang 平江

P'ing-t'ou 平頭

P'oyang 鄱陽

Prince Ch'ing 慶親王

Prince Ch'un 醇親王

Prince Kung 恭親王

P'u 鋪

p'u-ping 鋪兵

p'u-ssu 鋪司

P'u-ti 鋪遞

Sai-t'ou 賽頭

Samshui 三水

Sanyuan 三原

Sha-hu k'ou 殺虎口

Shanhua 善化

Shang-hai hsin-pao 上海新報

Shaohing (Shaohsing) 紹興

Shasi 沙市

Shen-pao 申報

Shen Pao-chen 沈葆楨

"Shen-pao kuan t'iao-li"
申報館條例

Shengking 盛京

shih-chiang hsüeh-shih 侍講學士

Shih-ch'iao 石橋

shih-lang 侍郎

Shih-t'ou 石頭

shou 壽

shou-pei 守備

"Shu-ch'ien chia" 數千家

Sian 西安

Siangtan 湘潭

Siangyang 襄陽

so 所

ta-ch'en 大臣

Ta-hsing 大興

Ta-huang 大黃

t'ai 臺

t'ai-chan 臺站

"T'ai-hsi yu-cheng k'ao" 泰西郵政考

Taiwan yu-p'iao 臺灣郵票

tan 石

t'ang 塘

T'ang chen 湯震

T'ang-pao 塘報

T'ao Mo 陶模

Tarbagatai 塔爾巴哈台

"Te chih liu-hsing su-yü chih-yu erh
ch'uan-ming" 德之流行
速於置郵而傳命

Te-tsung 德宗

t'i-t'ang kuan 提塘官

T'ien-kung 天宮

Ting-chia 丁家

Ting Pao-chen 丁寶楨

t'ing 廳

T'oung pao 通報

"Ts'ai hsi-hsüeh-i" 採西學議

"Ts'ai I-chan i" 裁驛站議

Tseng Chi-tse 曾紀澤

Tseng Chung-hsiang-kung shu-cha
曾忠襄公書札

Tseng Kuo-ch'üan 曾國荃

Tsinchow (Chinchow) 秦州

Tsinshih 津市

Tso Tsung-t'ang　左宗業

tsou-i　奏議

Tsou-shih ch'u　奏事處

Tsung-li ko-kuo shih-wu ya-men
　總理各國事務衙門

Ts'ung-lin　潀林

Tsunyi　遵義

Tu Mu　杜牧

Tu-shih k'ou　獨石口

Tung-an　東安

Tung-t'un　東屯

Tungchwan　東川

Tungchow　通州

T'ung-wen kuan　同文館

Turfan　吐魯番

Tz'u-an　慈安

Tz'u-hsi　慈禧

*Tz'u-yüan*　辭源

Uliassutai　烏里雅蘇臺

Urga　庫倫

Urumchi　烏魯木齊

Ush　烏什

Wa-tien　瓦店

Wai-pien　外編

Wan-p'ing　宛平

Wang I-san　王益三

Wang T'ao　王韜

Wanhsien　萬縣

wei-ju-liu　未入流

"Wei Ku mou fa Yung"　蒍賈謀伐庸

*Wei-yen*　危言

Wen-hsiang　文祥

Wen-pao chü　文報局

Wen-pao tsung-chü　文報總局

wen-pien　文編

Wen-tsung　文宗

Wo-jen　倭仁

Wu T'ing-fang　伍廷芳

Wuchang　武昌

Wuchengchen　吳城鎮

Wuchow　梧州

Wuhu　蕪湖

Wushan　巫山

Wusüeh　武穴

Ya-tzu　鴨子

Yang Kuei-fei　楊貴妃

Yang-lin　楊林

Yang Ping-chang　楊秉章

Yangchow　揚州

Yanghsien　洋縣

Yarkand　葉爾羌

Yin Hsien (Hung-ch'iao)　殷羡(洪喬)

Yochow　岳州

yu　郵

Yu-cheng　郵政

"Yu-cheng cheng-ts'e chih chien-t'ao"
　郵政政策之檢討

yu-cheng-chü　郵政局

Yu-cheng shang-p'iao　郵政商票

"Yu Chi shih-Ch'u"　游吉使楚

Yung　庸

Yung Wing　容閎

Yunyang　雲陽

# INDEX